THE MODEL RAILROADER'S GUIDE TO
INDUSTRIES 2
ALONG THE TRACKS

JEFF WILSON

KALMBACH BOOKS

Printed in the United States of America

10 09 08 07 06   1 2 3 4 5

Visit our Web site at
kalmbachbooks.com
Secure online ordering available

Publisher's Cataloging-In-Publication Data
(Prepared by The Donohue Group, Inc.)

Wilson, Jeff, 1964–
    The model railroader's guide to industries along the tracks. 2 / Jeff Wilson.

    p. : ill., map ; cm.

    Includes bibliographical references.
    ISBN-13: 978-0-89024-658-0
    ISBN-10: 0-89024-658-0

1. Railroads—Models. 2. Industrial buildings—Models. 3. Industrial equipment—Models. I. Title.
II. Title: Industries along the tracks

TF197 .W545 2006
625.1/9

CONTENTS

Acknowledgments.................................. 4

Chapter One: **Coal customers** 5

Chapter Two: **Milk and dairy traffic**....................... 19

Chapter Three: **Breweries** 31

Chapter Four: **Paper** 47

Chapter Five: **Iron ore** 59

Chapter Six: **Package and LCL traffic**..................... 75

Selected Bibliography 88

Acknowledgments

It's difficult to do justice to a complex industry in a single chapter of a book, but I've done my best to condense information and photos to that which I felt was of most use to modelers. The bibliography lists additional sources of information, which I encourage you to check out if you want more in-depth material.

A book like this would be impossible to put together without a great deal of help, and I would like to thank many people for their time, materials, and assistance. Several individuals spent considerable time going through their personal photo collections, and I thank them for sharing their images with me: Richard Cecil, Mike Danneman, Bob Gallegos, Jim Hediger, John Leopard, Mike Schafer, Ted Schnepf, Andy Sperandeo, Matt Van Hattem, and Hol Wagner. Thanks also to Neil Schultheiss, who allowed me to use an image from Boatnerd.com, an outstanding Web site that covers Great Lakes ships and shipping. Thanks also to Cody Grivno, Kent Johnson, David Popp, and other Kalmbach staff members for their brainstorming and ideas on photo and information sources.

Special thanks go to Dave Herrewig of Miller Brewing Co. for providing several outstanding historical images from that company's archives, to Cleo Rohrman of the Tecumseh Energy Center for taking me on a tour of that facility, to John Bromley of the Union Pacific and Stephen Priest of Paired Rail Productions for information and for their assistance in arranging visits to various industries, to pilot Brian Lenzen for giving me turbulence-free rides during aerial photography, to Peg Leinenkugel of the Leinenkugel Brewing Co. for providing information, and to Denise Zingg of Spectrum School of the Arts in Racine, Wis., who graciously let me use her school's darkroom.

The information in this book comes from hundreds of sources, including the books, magazine articles, historical society publications, company pamphlets, and Web sites listed in the bibliography, along with many other sources too numerous to mention. Conflicting information sometimes arose; I did my best to sort it out, and any mistakes that found their way into this book are mine and not those of others.

Coal customers

1-1

The first *Model Railroader's Guide to Industries Along the Tracks* examines how coal is mined and how railroads move coal. Mining is only the first stage of the coal business, so here we'll look at some major users of coal, along with modeling possibilities for different time periods.

Railroads themselves were once major coal customers, as steam locomotives burned 100 million tons of coal per year in steam's heyday. Today, electrical utilities are the country's largest coal customers, consuming about 80 percent of the one billion tons of coal mined each year in the United States, **1-1**.

Other customers over time have included steel mills, coke plants, coal gas plants (those that made gas from coal), small municipal power and heating plants, and local coal and fuel dealers. In addition, railroads haul coal destined for export.

Electric utilities are now the largest users of coal in the United States. This is the Tecumseh Energy Center near Topeka, Kan. *Jeff Wilson*

ONE

1-2 The Washington Terminal Co. Power Plant, built in 1907, supplied power for Washington Union Station. It was typical of small municipal power plants built across the country in the early 1900s. *Historic American Engineering Record*

Power plants

Today, 80 percent of mined coal is used to generate electricity. The total amount has doubled since the 1970s, when about 400 million tons were consumed each year.

As of the early 1990s, about 500 of the 3,000 power plants in the United States use coal. These range from small municipal power plants with capacities of 50 megawatts or less to large regional power plants generating 2,000 or more megawatts.

Power plants generally work on the same basic principle, with coal, oil, or gas firing boilers. The resulting high-pressure steam drives turbines that turn electricity-producing generators.

The first coal-fired power plant was built in 1882 in New York City, and from the 1890s through the early 1900s, communities built small power plants to provide electricity (or steam heat). These municipal power plants often had capacities under 100 kilowatts. Also common were power plants for industries, large structures, or complexes, **1-2**.

The boilers of most early plants were fired by coal or oil. Large diesel engines that drove generators directly, a system still used in many small plants, became practical in the early 1900s.

Following World War II many coal and oil plants were modified to burn natural gas, and many can use coal, oil, or gas, allowing utilities to choose the least expensive fuel at any given time.

Power plants, large and small, typically have more than one boiler and generator. This allows generators to be shut down during off-peak hours or seasons, and allows a generator or boiler to be maintained while others are up and running. For example, an early 2-megawatt municipal station might have three 500-kw and two 250-kw generators.

Muni plants come in many shapes and sizes, **1-3**. Constructed most commonly of brick, the

Municipal generating plants are found in a variety of shapes and sizes. Most, like this small plant in Fairfax, Minn., are built of brick and located next to railroad tracks. Few small plants are still served by rail. *Jeff Wilson*

At the rear of a municipal power plant, you'll find transformers, circuit breakers, and heavy wiring. High-voltage areas are always fenced. This is in Spencer, Iowa. *Ted Schnepf collection*

buildings are large enough to hold the boilers and generating equipment. Through the early 1900s, most generating plants were built along railroad tracks to provide easy access for fuel.

The front (street side) of these structures often resembles a standard office building, since they often included offices of the utility company. The track (rear) side of the building usually contains the boiler, generator, and other electrical equipment, and thus has fewer windows (or bricked-over windows). Tall smokestacks were the rule for early power plants: The lack of modern pollution-control equipment resulted in a lot of smoke, and a high stack was the best way to dissipate the smoke.

Some small power plants had fuel unloading areas inside the building. A spur like this would be an interesting model. *David P. Morgan Library collection*

1-6

Many older plants, such as the Tecumseh Energy Center, have been expanded over the years. The original brick part of the plant was built in the late 1920s and now serves only as office space. The coal unloading pit is under the shed (the plant switcher is visible through the shed), with part of the coal stockpile at right. *Jeff Wilson*

Outdoors at the rear, you'll find substation wires and transformers, **1-4**, and possibly storage tanks for fuel oil or diesel fuel next to a rail siding. Coal unloading was often done over a pit next to the building, where coal could be augered into the plant, **1-2**; it could also be unloaded over a trestle with a small conveyor system, or the coal could be offloaded inside the building, **1-5**.

When first built at the turn of the 20th century, many municipal plants were shut down overnight. By the 1910s, however, most were in 24-hour operation.

Through the World War I era and later, cities connected their municipal plants to power grids of surrounding areas, providing backup power in case of break-downs and, especially after World War II, giving communities a chance to buy cheaper electricity.

The number of small municipal plants began dropping considerably in the 1950s, as larger generating stations became more efficient. Also, many marginal small operators were financially unable to keep up with requirements for emission controls.

Today, many small muni plants (under 100 megawatts) still exist, but many only operate during peak times or serve as emergency backups. Few small plants still use coal, however, instead relying on natural gas or diesel engines to drive generators.

Rail operations at a small plant include incoming fuel loads. Coal arrives in hopper cars (not gondolas, since few small plants have rotary dumpers), and oil or diesel fuel is delivered in tank cars (don't forget the storage tanks). Depending upon the size of the plant, this could be one or a few cars at a time. Hopper cars or trucks could be used to haul away ash.

Plant sizes increase

Through the early 1900s, improved technology allowed power plants to continue to grow, and the 1920s and onward saw an increase in the number of larger plants that served multiple communities. When located away from the communities they served, heavy power lines carried their electricity. These plants were usually built to burn coal or oil.

Instead of small hand-fired boilers, these plants had larger automated boilers. During the 1920s, boilers first began using pulverized coal. In these boilers, coal ground to a powder is blown into the furnace, where it ignites and burns in suspension. Others use stoker furnaces, with crushed coal carried in on a moving grate.

Many plants built during the 1920s are still in service but have had new boilers, generators, and other equipment added to make them more efficient. These plants also have advanced air pollution controls to limit emissions.

The original portion of the Tecumseh plant, **1-6**, shows its 1920s heritage, typical of power plant architecture of the time, and is one that would be interesting to model. As the plant was modernized and expanded, new structures and equipment were added to the original building. Today, it houses only offices but remains an integral part of the plant.

Other equipment at power plants includes the usual collection of transformers, switches, and power lines, **1-7**; a cooling tower, **1-8**; holding ponds for water; and storage tanks for fuel oil if the plant can burn that fuel.

A prominent area of any coal-fired plant is the coal storage pile, **1-9**. Bottom-dump hopper unloading was the usual delivery method through the 1960s, so many older plants have a covered unloading area, **1-10**. The coal is dumped through the tracks to a receiving pit, where augers and conveyors carry it to the ready piles.

From the ready piles, conveyors carry coal into the plant to crushers that pulverize it for burning.

1-7

Power plants have substations with transformers, switches, circuit breakers, and power lines. High-voltage areas are fenced. *Jeff Wilson*

1-8

Cooling towers are prominent structures at power plants, especially in cooler weather when they send clouds of steam billowing skyward. *Jeff Wilson*

Power plants usually maintain at least a one- to two-month stockpile of coal to avoid any problems that might interrupt movement of coal to the plant.

Although the Tecumseh plant, **1-1** and **1-6,** receives coal in unit trains, the cars are only unloaded a few at a time. The plant receives a trainload of coal about once a week. Cars are left on spur tracks next to the main line, **1-11**, and brought into the plant by the plant switcher as needed.

Much of the coal for generating electricity once came from eastern mines, but the passage of the Clean Air Acts in 1970 and 1990

1-9 Large conveyors carry coal from the ready piles to the power plant. Several bulldozers move coal around the piles. *Jeff Wilson*

1-10 Coal is unloaded from bottom-dump hopper cars into a pit. Conveyors then carry the coal to the ready piles. *Jeff Wilson*

1-11 Loaded hoppers are left on a spur next to the main line. They're brought into the plant five at a time by the plant's 44-ton switcher. *Jeff Wilson*

has forced many utilities to choose western coal, which has a lower heating value but a much lower sulfur content. When burned, sulfur in coal produces sulfur dioxide, a major air pollutant. High sulfur content also accelerates corrosion of boilers, another reason for utilities to choose low-sulfur coal.

Large modern plants

During the 1950s and 1960s, larger power plants were built to run on oil or gas, but rising oil prices and the oil crisis of the early 1970s led to an increase in coal usage for power plants. Building large plants or converting them to coal led to the widespread use of unit trains for hauling coal.

Into the 1960s, coal bound for power plants was largely treated as any other commodity. Bigger plants would receive large cuts of cars, with smaller shipments of single cars to smaller power plants. As plant size increased, utilities realized the viability of buying coal a trainload at a time. Carrying this a step further, many utilities began buying or leasing their own coal cars.

Modern coal-burning power plants have varying capacities, but a typical plant, **1-12**, may have a 1,200-megawatt capacity, while some larger plants are nearly

1-12

The WE Energies Pleasant Prairie power plant near Kenosha, Wis., is a typical modern power plant. It receives unit trains of coal and has a loop of track surrounding the plant. *Jeff Wilson*

double that capacity. Modern plants have advanced coal-dust recovery systems along conveyor lines and unloading areas as well as modern electrostatic precipitators and other equipment to remove ash and other pollutants from emissions.

In modern plants, **1-12**, rotary dumpers unload cars, meaning large bathtub-type gondolas can be used instead of hopper cars. Early rotary dumpers could only

handle one car at a time, requiring cars to be uncoupled and switched into position, **1-13**.

The 1970s saw the development of coal hoppers and gondolas with rotary couplers at one end, allowing them to stay coupled during unloading, **1-14**. As the car passes through the dumper, clamps hold the car to the rails, and the car is rotated, centered on the coupler. It takes about 20 seconds for the coal to clear the

car. The car is then rotated back, the train moved forward, and the process repeated. Dumping takes about two minutes per cycle.

Unloading usually takes place in a covered building. In colder regions, the building might be longer to accommodate a car heater on the inbound side of the loader to thaw frozen loads. Modern high-tech unloaders, with computerized controls, allow a single operator to run the

1-13

Early rotary dumpers required cars to be switched into them one at a time. *Linn H. Westcott*

unloader. The cars are guided into position by an indexer, **1-15**.

To speed unloading, most plants have a continuous track loop, allowing the train to stay together during unloading. When unit trains are delivered to a power plant with continuous unloading capability, there's usually a time incentive for quick unloading, since the locomotives and crew stay with the train. It takes about four to five hours to unload a 110-car train.

The amount of coal consumed at a plant varies by its size, whether other fuels are being used, whether the plant is working at full capacity, and by the coal's heating qualities. A rule of thumb is that one ton of coal produces about 2,000 kilowatt-hours of energy. A large 1,200-megawatt plant will burn about 16,000 tons a day (just over one trainload), while a small (250-megawatt) plant that also uses gas might only require a train once a week.

In addition to inbound coal loads and outbound empties,

power plant rail operations may also include handling ash, a by-product of burning coal. Some of this ash is used as filler in concrete products; the remainder goes to landfills either on- or off-site. Much of the ash going to landfills is hauled by truck, but the ash heading to processing is often hauled by covered hopper.

Transloading operations often take place at plants, **1-16**, that contract with concrete companies for their ash. Ash is loaded at the plant into a pressurized truck, hauled to trackside, and transloaded into pressurized hoppers for transport to a concrete manufacturer. Some plants load the ash directly into railcars from overhead storage bins.

Coal dealers
Through World War II, coal was a popular choice for home heating and cooking. Every town had at least one coal dealer (almost always served by rail), and large cities had several. Some of these businesses only sold coal; others

1-14

Clamps hold the car in place as this modern bathtub gondola goes for a spin while still connected to the rest of its train. This unloader is at the Superior Midwest Energy Terminal coal dock in Superior, Wis., but it is typical of modern dumpers at power plants and other industries. *Boatnerd.com*

sold lumber, oil, gas, hardware, and other products.

These dealers came in many shapes and sizes, and many make attractive subjects for modeling. A typical arrangement for a small dealer, **1-17**, includes a rail siding elevated on a trestle. This allowed easy unloading of the coal to pockets under the trestle. Various grades of coal and coke could then be divided in these pockets.

This large computer-controlled indexer, located in the unloading shed at the SMET facility, properly positions the car on the rotary dumper for unloading. *Matt Van Hattem*

1-16

At the Tecumseh plant, ash is transloaded from a bulk semitrailer to a covered hopper for shipment to a concrete plant. *Jeff Wilson*

1-17

The Herbert E. Miller Fuel Co. is a typical small coal dealer, with a delivery trestle, several piles of various types of coal and coke, and delivery trucks. *B. F. Marteu*

1-18

1-19

Some coal trestles were covered. This one is covered by an extension of a building's roof. Note the loading conveyor at right. *Jeff Madden*

Larger coal dealers had tall storage silos. Cars unloaded into pits next to the silos, then the coal was conveyed to the enclosure above the silos. *David P. Morgan Library collection*

1-20

1-21

This large coal dealer along the Maryland & Pennsylvania (left) has an enclosed storage building, with truck-loading spouts coming down from each storage bin. *David P. Morgan Library collection*

The Chesapeake & Ohio docks at Presque Isle (Toledo), Ohio, are typical of rail-to-ship transloading docks through the early diesel era. The car at left is about to go for a ride up the inclined track. *Chesapeake & Ohio*

These trestles were sometimes covered, **1-18**.

Coal was loaded onto a dealer's truck using portable conveyors. The dealer drove to the waiting coal chutes of businesses and homes (a coal truck delivering a load makes an interesting model).

Larger dealers had storage silos or bins, **1-19**, that could each hold different types of coal. A vertical conveyor carried the coal from hoppers under the tracks to the top of the bins, where another

conveyor transferred coal to the appropriate bin.

In an arrangement for a larger dealer, **1-20**, covered elevated bins held the coal, and chutes dropped down from the bins to load trucks, eliminating the need for portable conveyors.

Coal, especially anthracite for home use, was often branded. Companies dyed some of their coal to distinguish it from other brands. Blue Coal was a popular brand of anthracite (you can see a

Blue Coal advertising sign on the end of the coal building, **1-20**).

As more and more households and businesses turned to oil and natural gas for heating, the number of coal dealers dropped from the 1940s onward, and coal dealers became difficult to find by the end of the 1960s.

Steel mill and export traffic

A significant amount of coal hauled from the mid-continent and Appalachian areas was

1-22

A cable-powered ram, which travels in a gap between the rails, shoves the loaded car up the ramp. *William A. Akin*

1-23

Once at the top, the car is clamped to the rails and turned over. Shown here is the Pennsylvania Railroad at Sandusky, Ohio, in 1943. *Library of Congress (call no. LC-USW361-661)*

1-24

Once the car is unloaded, it will roll down from the dumper, through a spring switch, and up the kickback (foreground). It will then roll down the track at left and back toward the yard. Multiple dumpers are in operation on most docks. *David P. Morgan Library collection*

destined for steel mills along the Great Lakes. Much of this coal was (and still is) loaded onto ships or barges for delivery. Major shipping points included Toledo, Ohio, and Superior, Wis.

About half the exported Appalachian coal goes out of Newport News, Va.; other significant export docks are at New Orleans, Cleveland, and Baltimore.

Transloading docks are fascinating in both appearance and operations, **1-21**. Cars would be taken up a ramp one or two at a time to dump their loads directly into a ship. The cars were raised to the top of the dock by a cable-powered ram, often called a pig or a mule, **1-22**. Later docks used a side-arm guide on a parallel track to shove cars up the grade.

Clamps inside a dumper secure the car, **1-23**, as the dumper rotates the car, emptying the coal either into holding bins or directly into a ship's hold. The dumper then turns the car upright.

Gravity then takes over. The car is rolled out the far side of the dumper, through a spring switch, and up a steep kickback, **1-24**. The car then rolls down off the kickback, **1-25**, through the switch, and down to a collection track. Switch engines then collect the empties and pull them back to a storage yard, from where they are sent back to the mines.

The staging and storage yards for these docks are huge, **1-26**. Incoming trains are sorted by grade and size of coal. Locomotives then shove the appropriate cars to the docks as needed.

Here's an idea of the volume of coal handled this way: In 1950 at Toledo, the Chesapeake & Ohio unloaded 9.5 million tons of coal; the Norfolk & Western, 2.6 million tons; the Louisville & Nashville, 2.5 million tons; and other railroads combined for 2 million tons. That's more than a quarter million hopper cars passing through that city's docks alone.

Great Lakes coal numbers dropped in the 1960s, with the amount loaded at Toledo falling to about 5 million tons per year in

1-25

An empty car rolls up the kickback. It will then roll back down through the spring switch and be added to a train of empties bound for the coal fields. *William A. Akin*

1-26

The staging yards for coal docks are extensive, as shown by the New York Central's yard at Ashtabula, Ohio. Note various sizes of coal in the hopper cars and the many styles of cars visible in this early 1940s scene. *New York Central*

the mid-1990s. The shift to low-sulfur western coal made Lake Superior the leading coal-loading point on the Great Lakes, at more than 13 million tons per year.

Modern docks, like the Superior (Wisconsin) Midwest Energy Terminal, **1-27** and **1-28**, can unload solid unit trains of coal using rotary dumpers, **1-14** and **1-15**. Conveyors then deliver the coal to the holds of ships.

Steel mills also often receive coal and coke by rail, **1-29**. For more information on steel mills and their operations, see the book *The History, Making, and Modeling of Steel* by Dean Freytag.

Railroad coaling towers

Through the 1940s, railroads consumed around 100 million tons of coal per year. Most of this went to steam locomotives, so don't forget to include coaling towers as customers for coal. Most have unloading pits next to them, so hoppers can be spotted, **1-30**.

The amount of coal used depends upon the capacity of the tower and how frequently locomotives take on coal at the tower. By adding the tender capacities of the steam locomotives that are serviced daily, you can get a good

1-27

A Burlington Northern Santa Fe train rolls slowly around the unloading track loop at the Superior Midwest Energy Terminal (SMET) transloading center in Superior, Wis., in 2003. The ship at left is the *Columbia Star,* a thousand-foot Great Lakes boat with a capacity of 78,850 tons. *Matt Van Hattem*

idea of how many coal hoppers per day you'll need to keep your locomotives supplied.

Coal cars

The 50- or 55-ton, two-bay hopper car, **1-26**, was the most common coal carrier from World War I through the end of the steam era, and it remained common into the 1970s. The next larger car, the 70-

ton three-bay hopper, **1-30**, came into wide use in the 1940s, and it was common into the 1980s.

Some railroads used larger (100-ton) hoppers and gondolas early in the 20th century, but they didn't become common in interchange traffic until the 1960s with the coming of the coal unit train, **1-29**. Today 100- and 110-ton bathtub gondolas, **1-15**, and

At the SMET terminal, conveyors move coal from the unloading area to large storage piles, then from storage piles to ships. *Matt Van Hattem*

Trains often deliver coal directly to steel mills. *Jim Hediger*

Several hopper cars are spotted on the unloading track at this Ontario Northland coaling tower in Cochrane, Ontario. *Mike Runey*

hoppers are the standard for unit trains, with many older 100-ton hoppers still in service for single-car shipments.

Modeling

Walthers has offered a silo-style coal dealer, Goldenflame Fuel Co., in HO (no. 933-3087) and N scale (933-3246), a trestle-style dealer in HO (O.L. King, 933-3015), and coal conveyors, a coal truck, and other details. The N Scale Architect makes an HO kit for a silo-style dealer (no. 40004).

If you're modeling a power plant, Walthers has made modern (HO no. 933-3055) and older brick-style power plants (HO, 933-3021; N, 933-3214), along with transformers, substations, and other accessories, and DPM has a brick power plant in HO (no. 356). Many brick-structure kits would work fine in making a smaller municipal power plant or industry power house.

Many companies offer hopper cars and gondolas, and coal loads are made by Accurail, Chooch, and others. Other details include vehicles (Jordan makes an old Mack coal truck in HO) and scale coal, available in various grades from Noch, Highball, Woodland Scenics, and others.

Coal dealers and small municipal power plants are the easiest to model. Both are visually interesting and help place your layout in an era, and many can be served by a single rail spur. Modeling a grade separation for a coal trestle would also be interesting.

Few layouts have the space needed to model a large power plant with a loop track, but you might be able to do it by having the track disappear through a backdrop, possibly by disguising the hole in the backdrop as the entry to the rotary dumper.

It is easier to model a smaller power plant, where a local train delivers a cut of cars (or an entire train) to a spur next to the main line, than have the plant switcher – or a local freight – move a few cars at a time into the plant to the dumping shed.

Even if you don't model an on-line industry, you can model the trains serving them. You can have solid unit trains passing through (with run-through locomotives from a different railroad, if you like).

Or you can have trains with single cars or cuts of multiple cars passing through on their way to a coal dealer or muni power plant, with loads heading one direction and empties going the other way.

Milk and dairy traffic

2-1

Borden's Creamery, Cochecton, N. Y.

Milk was once a high-priority commodity for railroads. Milk traffic and dairy operations are fascinating and lend themselves to modeling. Northeastern railroads had dedicated trains carrying milk from farms and rural creameries, **2-1**, to large creameries and processing plants in cities such as Boston, New York, and Philadelphia. Other products of larger creameries, such as evaporated and condensed milk, cheese, and butter, shipped by rail throughout the country.

Railroad milk operations peaked in the late 1920s and early 1930s. Although some lines hauled milk through the 1960s, most traffic moved to trucks by the end of the 1950s due to improved highways and higher capacity tank trucks. Trains still haul processed dairy products, such as condensed and evaporated milk, milk protein, and cattle feed additives.

The Borden creamery in this postcard view is typical of the wooden-frame creameries found throughout the Northeast. The scene is along the Erie RR at Cochecton, N.Y., in the early 1900s. *Jeff Wilson collection*

TWO

2-2

Rutland train no. 8 carries a variety of milk cars, including a couple of National Car Co. flatcars with milk tanks and a steel Pfaudler car. *Jim Shaughnessy*

History

Farming from the 1800s into the early 1900s was much different than today. With no electricity for automatic milking machines or refrigeration, farms were necessarily small, with just a few cows milked by hand every evening and morning. Through the mid-1800s, most of this milk was used by farmers themselves or traded or sold locally.

Creameries began to appear in the early 1800s, serving as milk-collection points for farmers. Farmers put milk into cans after the evening and morning milkings. The evening cans had to be kept cool overnight, so they were placed in tubs or vats of cold water, either supplied by a spring (in the spring house) or by ice obtained from the creamery.

As soon as the morning milking was done, the farmer brought his cans to the local creamery. There, the milk would be consolidated with that from other farmers, and either processed into products such as butter or cheese, or shipped to a large city creamery.

Milk shipments by rail began in the 1840s, but didn't grow to a significant volume until after the Civil War. Milk traffic was most evident in New England and the Northeast, with New York and Boston receiving large amounts via several railroads. Other regions of the country saw milk traffic as well, including the Chicago area, Ohio, Michigan, and Southern California.

Improved rail service into the 1900s enabled railroads to ship milk to cities from creameries farther and farther from the big city. Railroads hauling substantial milk traffic included Bangor & Aroostook; Boston & Maine; Canadian Pacific; Central of New Jersey; Central Vermont; Delaware & Hudson; Delaware, Lackawanna & Western; Erie; Lehigh Valley; Maine Central; New York Central; New York, New Haven & Hartford; New York, Ontario & Western; and Rutland.

These railroads operated dedicated milk trains, which hauled their own milk cars as well as those of creamery companies, **2-2**. Many other railroads throughout the country hauled milk as well, but the volume didn't warrant dedicated milk cars or trains. In these cases, milk cans were hauled as express in standard baggage cars and trucked from the station to a creamery at the destination station (see photos **6-2** and **6-23** in Chapter 6).

Milk was only one dairy product that moved by rail. Butter and cheese, with longer shelf life, were shipped by rail across the country, **2-3**, along with condensed, evaporated, and dry milk.

By the end of the 1950s, most milk traffic had moved to bulk tank trucks. The Maine Central carried milk cars into 1969, and the last rail-carried bulk milk was hauled into Boston on the B&M in 1972.

Creameries

A creamery can serve many purposes. Its basic function is as a collection point for milk from local dairy farmers. Many creameries are cooperatives, including large organizations such as Dairymen's League and Land O'Lakes. Other creameries were, and are, owned by large dairy companies such as Borden or Hood.

Many early creameries were small operations, and into the early 1900s, creameries were located close together – they needed to be just a short wagon trip away from the farms they served. With the coming of motor vehicles and improved roads, many smaller creameries went out

2-3

Cases of butter – one-pound cartons packed in larger cardboard cases – are loaded into wood-sheathed ice-bunker refrigerator cars at the Land O'Lakes plant in Minneapolis, Minn., in 1941. *John Vachon, Library of Congress (call no. LC-USF34-063410-D)*

of business, and the remaining creameries grew larger.

Creameries varied in style by region of the country. In New England, most were wood, with some block or brick, **2-4**; in the Midwest, they tended to be built of brick, **2-5**. Creameries that shipped milk by rail had platforms at rail level. Most had sidings sized to hold a single railcar, although some could load two cars.

Creameries had tall smokestacks for their boilers. A large area was needed for bulk storage tanks or for ice-water vats that stored milk cans. Until the development of mechanical refrigeration in the early 1900s, most creameries had ice houses, either attached or as separate buildings.

Farmers would pull up to the creamery, then place their milk cans on conveyors that carried them through a can door, **2-6**. Empty cans would then come out another can door, usually around the corner from the entry door.

2-4

A railroad-owned can car (middle) and a standard 40-foot boxcar rest at the New England Dairies creamery on the Central Vermont Ry. in Enosburg Falls, Vt., in 1941. The truck at left is unloading cans of milk collected from dairy farms. *Jack Delano, Library of Congress (call no. LC-USF34-045973-D)*

2-5

Midwestern creameries, like this rail-served one in Winthrop, Minn., were often made of brick. *Jeff Wilson collection*

Workers move cans from a truck onto a conveyor, which takes the cans into the creamery. *Linn H. Westcott*

Once the cans are inside the creamery, workers weigh and test the milk. The cans are then steam cleaned. *New York Central*

Farmers generally brought their own cans to creameries (often a single can or just a few) throughout horse-and-wagon days, but in the 1920s and '30s, many creameries acquired trucks to make the rounds of local farms. Farmers would often leave their cans next to the road (sometimes on a small platform) to speed the operation. Bulk trucks for farm pickup began in some locations in the 1930s. Some of these bulk trucks carried milk to the local creamery, some directly to large city creameries.

As milk cans arrived, they were weighed (payment was by weight, not volume) and tested for temperature, odor, bacteria, and butterfat content, **2-7**. Grade A milk was destined to be sold as milk; Grade B milk would be made into cheese or other products. Milk would either be placed into bulk tanks or into the creamery's own milk cans and placed into cooling vats. Some creameries separated cream at this point and dealt with it separately.

Once emptied, the farmer's cans were either returned directly to him via a second can door, whereupon he would take them home and clean them; or the cans would be run through a steam cleaner, then returned.

What happened next depended upon the creamery. Smaller collection creameries immediately shipped the collected milk by rail or truck to larger city creameries. Some creameries bottled milk for local sales and for shipment to the city (in cases), shipping out the remainder of the milk in cans or bulk. Other creameries specialized in making butter or cheese and shipped out only the excess milk not needed for their processing.

Creameries that bottled required sufficient space for the bottling and pasteurization equipment as well as storage for bottles and cases. Bottled milk was the standard through the 1940s, when cardboard cartons became more common.

Into the early 1900s, some farmers without easy access to creameries instead used rail-side platforms. Farmers would drop off their cans at the platform prior to the train's arrival, whereupon the cans would be loaded into a can car. These platforms were usually located directly on the main line to save time, since the stop only took a few minutes. These stands pretty much disappeared by the 1920s as trucks were used more and more for picking up cans.

Changes in the creamery industry came with the widespread use of mechanical refrigeration through the 1930s. The technology was too expensive for many small creameries, so they closed, transferring their business to other nearby creameries, which in turn grew. This gave railroads fewer shipping points. As an example, in 1930, Dairymen's League had 273 country creameries; by 1936 this number had dropped to 117.

Other dairy plants

Butter became a major dairy product in the late 1800s, in part because it had a fairly long shelf life compared to milk. Early creameries would pack butter into large wooden tubs, but by the early 1900s, butter was being packaged for home use, mainly in one-pound packages.

By the 1940s, butter plants were packing tubs destined for food processing plants and other bulk customers, such as the military, **2-8**, as well as filling cardboard cases of packaged butter for consumer use. Many of these creameries were large operations, **2-9**.

Condensed and evaporated milk are made at creameries known as condensaries, **2-10**. Gail Borden patented a method for condensing milk by adding sugar to the water from milk and boiling it at low heat in a vacuum. He opened his first condensary in 1857, marketing Eagle Brand Condensed Milk. It had a long shelf life, which was a big selling point in the days before home refrigeration. The product was packed in tin cans and shipped by rail throughout the country.

A related product, evaporated milk, was developed later. Evaporated milk is basically condensed milk without the sugar. Condensed milk became an important ingredient in baking, ice cream, and candy, while evaporated milk was used for making milk. By the late 1800s, Borden had licensed other plants to make the product, and other companies, such as Carnation and Pet, began making these products across New England and the Midwest, as well as other areas of the country.

Condensaries were larger operations than creameries. They generally didn't ship milk in bulk, instead making up carloads of their products. Although evaporated and condensed milk didn't require refrigeration, they could be shipped in refrigerator cars (without ice) or in standard boxcars.

Dried milk was yet another product often shipped by rail. Dried milk was usually made from the skim milk that was a by-

Butter was often packed in tubs for bulk users. This is the Land O'Lakes Minneapolis plant in 1941. *John Vachon, Library of Congress (call no. LC-USF34-063383-D)*

Large-city creameries were big operations. This Land O'Lakes plant in Minneapolis, shown in 1929, produced butter, dried milk, and many other products. *Minnesota Historical Society*

Condensaries were generally large operations. This 1947 view along the Monon shows the Wilson's Milk plant of the Indiana Condensed Milk Co. at Sheridan, Ind. *Linn H. Westcott*

2-11

Wood barrels of dried milk await shipping at an Antigo, Wis., creamery in 1941. *John Vachon, Library of Congress (call no. LC-USF34-063454-D)*

2-12

The Kraft-Phenix cheese plant in Beaver Dam, Wis., shipped by rail and truck in the 1940s. *Jeff Wilson collection*

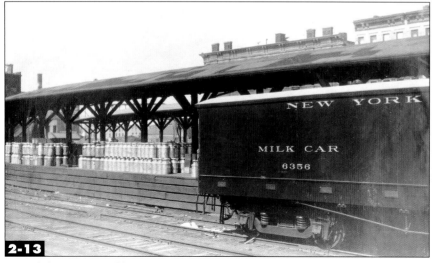

2-13

Large platforms were used for transloading milk cans from railroad cars to trucks in New York City. The New York Central milk can car resembles a wood express refrigerator car. *New York Central*

product of buttermaking. Into the early 1900s, the only use for this skim milk was as feed for calves and pigs. By the 1920s, candy companies began using large amounts of skim milk for making chocolate. Other dried milk products, including lactose and whey protein, were used in cattle feed.

Some skim milk was shipped in bulk, but drying the milk to powder was a more cost-effective way of shipping it. Through the 1940s, it was often packed in wood barrels, **2-11**; later it was shipped in bags and sacks.

Cheese plants were another specialized operation. Because it kept longer than raw milk, processed cheese of various types became popular in the early 1900s. One of these, Philadelphia Cream Cheese, made by Phenix, had been popular since the 1880s, but processes extended the product's shelf life from a few days to 120 days, giving it a national market.

Kraft was a major producer of processed cheese products from the early 1900s, producing about 40 percent of U.S. cheese products when it merged with Phenix in 1928. Among many other products, the company introduced Velveeta in 1928, boxed macaroni and cheese in 1937, and boxed cheese slices in the 1940s. These products were, and are, made in plants throughout the country, and shipped nationwide by rail and truck, **2-12**.

Can cars

In the early days of milk operations, railroads simply hauled milk cans from rural creameries to the city in standard baggage cars. This continued through the mid-1900s in many areas where milk traffic was low and didn't warrant specialized cars.

However, in New England as milk traffic increased in the late 1800s, railroads began dedicating cars to hauling milk. The style of these railroad-owned cars varied by railroad, with many early cars rebuilt from older baggage cars.

Milk cars evolved in the early 1900s, with most resembling express refrigerator cars. Although some were refrigerated, most had insulated bodies and swinging doors but without ice bunkers or roof hatches, **2-4** and **2-13**. Other milk cars looked more like standard refrigerator cars, but again without the roof hatches. This was because the milk was already cold when it was loaded, and the car's insulation was usually enough to keep milk cool for the relatively short trip. Most railroad-owned cars were wood, but some railroads had steel milk cars – notably the Erie and Boston & Maine, **2-14**.

Since they were considered head-end cars (see the sidebar on page 28) and were often hauled in passenger trains, milk cars rode on express or other high-speed trucks, and were equipped with steam and signal lines.

Railroad-owned milk cars were generally painted in each railroad's standard passenger train colors, usually Pullman Green. They also usually were lettered for their service.

Known simply as can cars, they carried milk cans in two layers. Some cars had fold-down racks for the second layer of cans; in other cars, cans would just be stacked atop each other, with belts or nailed-on brackets holding them in place. Cars traveling to the city would also sometimes carry cases of bottles or cartons.

For cans traveling longer distances, ice would sometimes be shoveled directly on top of the cans. Some railroads also used insulating blankets over the cans.

Into the 1920s, railroad-owned can cars were the usual way of getting milk to the city. However, handling cans was labor intensive, and returning empty cans was inefficient. As milk traffic increased, railroads and creamer-

2-14

Boston & Maine 1934 is one of 20 insulated steel cars built for can service in 1957. Another 15 similar cars had two doors per side and were equipped with mechanical refrigeration for hauling cases of bottled milk. *Scott Hartley*

Milk can

The 10-gallon milk can is an icon of the dairy industry. Many other sizes were used at one time or another (smaller cans were often used for cream), but the industry had pretty much settled on the standard 10-gallon (40-quart) size by the late 1800s. When full, one weighed about 100 pounds – small and light enough for an average man to be able to handle without trouble. Cans came in several styles, with various handles and lids, but all measured about 14" across and 24" tall.

In early days, farmers generally owned their own cans for transport to creameries, and creameries owned cans used for further transportation, but as farms and creameries grew, creameries also owned the cans used by farmers. Cans were usually identified by small plates affixed near the top of the can, and they might have painted labels or numbers as well.

ies looked to improve milk handling.

Bulk milk cars

The answer was the milk tank car. Among the first were three cars built by General American in 1922 for Wieland Dairy of Chicago. On the outside, the car looked like a refrigerator car without ice hatches, but inside was a pair of 3,000-gallon, glass-lined tanks built by Pfaudler, a company that

made tanks for the brewing industry.

By the late 1920s, General American was building production cars, and in 1930, General American and Pfaudler formed General American-Pfaudler, leasing and selling bulk milk cars to several private owners. (Individual railroads didn't own these cars.) Early production cars had wood sides, were 40- and 50-feet long, and were built for passenger

2-15

Sheffield Farms owned this early General American-Pfaudler wood tank car, shown here on the New York, Ontario & Western in 1940. Large sign boards on milk cars were common into the 1940s. *R.O. Hardy*

2-16

Workers connect a hose to unload milk from a Dairymen's League 50-foot steel Pfaudler car into a semitrailer in New York City. Team tracks were used to transfer loads to off-line creameries in larger cities. *New York Central*

service with express trucks and steam and signal lines, **2-15**.

Steel cars, also in 40- and 50-foot lengths, began appearing in the late 1930s. Early cars had double swing doors, while later cars had single swinging doors. Pfaudler cars had a small sliding door above the entry door, sometimes used for the unloading hose, **2-16**. The cars also had electric lights and an electrically powered agitating mechanism for the tanks, which was plugged in while the cars were loading at creameries. Cars used either a pump or air compressor for unloading.

General American-Pfaudler was the most prolific builder, but others, notably MDT, built cars in the late 1920s using Pfaudler tanks. Borden and Whiting were among the owners of these cars, which could be spotted by their C-channel sideframes.

Probably the most famous of the milk cars was Borden's so-called butter dish car, **2-17**. These cars, rebuilt from earlier wood cars starting in 1935, had pairs of 3,000-gallon tanks covered by a shroud made of formed steel with an aluminum coating. They originally had fins along the top

and ends, which were removed during World War II. The cars' advantage was their lower weight compared to a house car. General American-Pfaudler built similar cars for Hood's as well, **2-18**.

Bulk cars were originally precooled by pumping cold brine through piping inside the car at the loading creamery. This was found to be unnecessary, and most cars had this equipment removed in the 1930s and '40s.

As cars were loaded at a creamery, it was important to spot cars properly so that pipe connections could be made, **2-19** and **2-20**. Cars were unloaded at the city creamery or more often at a team track or terminal by hose into bulk tank trucks, **2-16** and **2-17**.

All milk tank cars were privately owned. Through the 1930s, these cars often had their owners' or lessors' names on large signs on the sides, **2-16**. Later, graphics became more reserved, with smaller lettering.

Tank containers

Since many metro creameries were located off-line, getting the milk from railcars to the creameries was time and labor intensive. An innovation to speed operations was a detachable milk tank that could be transferred from flatcar to truck chassis.

The first of these tanks had a 2,000-gallon capacity with an oval cross-section and sat in a steel cradle with runners on the bottom. They were loaded perpendicular to their specially equipped flatcars, and one person could transfer the tank to a truck in about 90 seconds.

At first, flatcars could carry three tanks; later ones could haul four. As with bulk cars, flatcars were equipped with high-speed trucks and steam and signal lines for passenger-train service.

Later tanks were larger – 2,500 and 3,000 gallons – and had a modified shape with a wide, flat base. They eventually used a side-

2-17

Borden's butter dish cars were the most-recognizable milk tank cars. Here milk is transferred from one into a truck at New York. *New York Central*

to-side transfer mechanism, so they rode lengthwise two to a flatcar, **2-2**. By the 1940s, National Car Co. operated more than 100 cars (with NX reporting marks).

These tanks were transloaded only at the unloading end of their journey. Rural creameries loaded the tanks while on flatcars, just as with other bulk cars.

Trailer-on-flatcar tank service was also tried, but the practice wasn't widespread.

Bulk cars became the primary means of hauling milk and continued to be so into the 1940s, although some creameries still shipped cans through the 1940s. Most milk trains carried a mix of car types, **2-2**. The mix of can cars, bulk cars, and containers varied by railroad and by the companies served by each.

Train and terminal operations

By the early 1900s, milk had become a big revenue source for many railroads. It was a profitable

2-18

A St. Johnsbury & Lamoille County local sets out a pair of empty milk cars at a Hood's creamery in Cambridge Junction, Vt., in 1946. Both cars were built by General American-Pfaudler: a wood tank car and a hooded tank car similar to a Borden car. *Philip R. Hastings*

operation, and trains hauling milk were given high priority – in fact, many milk trains operated as part of passenger trains or ran on passenger schedules.

Milk train schedules had to be relatively fast to get the milk to the city before it got warm and spoiled. Although milk trains

made many stops and had what appeared to be slow, lazy schedules, they kept moving.

Specific operational details varied by railroad, but most routes that handled high milk volumes followed the same basic pattern. The milk train would begin its run to the big city in the

2-19

A Borden car is loaded at a small-town creamery. Piping connects the car to the creamery, meaning that the car must be spotted at precisely the right place. *New York Central*

Milk car classes

The Association of American Railroads divided cars into classes depending upon equipment and loads handled. These classes can be useful when looking up information, especially in Railway Equipment Registers. Here's a summary of car classes used for hauling milk:

BM—A non-refrigerated car equipped for passenger service and used primarily for transporting milk in cans or bottles (such as most railroad-owned can cars).

BMR—Same as BM, but insulated with ice bunkers or ice boxes.

BMT—A non-refrigerated house car equipped with one or more insulated tanks (most privately owned milk cars, including Pfaudler cars).

BE—Baggage express cars (sometimes used in milk service).

BR—Refrigerator express cars (sometimes used in milk service).

farthest outlying town with a creamery, which depending upon the route, railroad, and era could be up to 400 miles from the city.

A train would leave early in the morning, once milk was collected at the farthest outlying creamery. The train would stop at each creamery or can loading platform along the line, **2-21**, loading and picking up can cars and picking

up bulk cars – from several companies – as it went, arriving at the city terminal by evening.

Empty cars from the previous day would be ready for pickup at the terminal (if they hadn't already been picked up by a train earlier in the day). Empty cars were collected – along with empty milk cans – and the return train would head out, dropping off

empty cars as needed along the line as it headed back to the originating town.

Some railroads would have multiple trains carrying milk traffic, depending upon the volume and number of creameries along various stretches of railroad, **2-22**. Trains would also make connections from different divisions or branch lines. Late arrivals at junctions might mean handling milk in a later train.

Some cars were delivered directly to creameries in the big cities. An example was the Hood's creamery in Charlestown (Boston), which had three tracks holding 15 cars and saw multiple switch jobs each day.

Much more common, however, was off-loading at a terminal or team track, since many city creameries were located off-line. Here, cans were off-loaded onto platforms and loaded into trucks, **2-13**, milk containers were off-loaded to trucks, and bulk tanks transferred their loads to tank trucks, **2-16** and **2-17**.

Each railroad and route had a unique mix of creameries owned by various companies or cooperatives along each line. Private-owner milk cars would indicate the destination of a train, since not all creamery companies operated in multiple cities. For example, New York creameries included Borden's, Dairymen's League, and Sheffield; Boston creameries included Whiting and Hood's; and Philadelphia had Supplee.

Switching operations

Individual train operations at each creamery involved setting out or picking up one to two cars. Double-ended tracks were common at creameries, greatly speeding operations since loads and empties were handled in opposite directions. Milk cars were usually carried at the head end of trains, which simplified switching. Empty cars were sometimes carried at the ends.

At small creameries, cans would often be added to a can car already in the train, while a larger creamery would have its own can car spotted the day before. Can cars were simply spotted at loading docks. Creamery company employees, called handlers, would ride trains and take care of loading and unloading milk cans.

Tank cars required careful spotting, since the piping used to load the cars had to be precisely aligned with the car door, **2-19** and **2-20**. Bulk cars were generally picked up fully loaded at creameries, although occasionally a car would be loaded at two different creameries.

Privately owned (and leased) cars would only be spotted at their owners' creameries. In other words, a Borden car would always be spotted at a Borden creamery, never at a Sheffield creamery.

Other operational challenges included timing train arrivals with farmers and delivery trucks arriving at the creamery, as well as making connections with other trains.

Condensaries, cheese plants, and butter plants would sometimes ship out cars of milk or skim milk, but carloads of cases of products in standard freight cars were the main outbound load. Since these products weren't destined for city terminals and weren't time-sensitive, they would generally be switched by way freights and handled in standard freight trains.

These plants also often received inbound loads, including crates, cardboard boxes, containers, and other materials.

Many plants still ship or receive products by rail, **2-23**, including dried protein, whey, cheese, and various milk products.

Modeling

The most obvious way to model the dairy industry is to duplicate the operations of a milk train, having it serve a creamery or two

A worker tightens the connections on the loading pipe, which extends through the tank car's access door. *New York Central*

The Hood's creamery at Sheldon Junction, Vt., had its siding on a grade. To pick up cars, the train is broken, then the car's brakes are released and it rolls into place. The car is a Pfaudler steel car. *Philip R. Hastings*

Boston & Maine train no. 5507 ran daily except Sunday from Boston to Bellows Falls, Vt., carrying milk and express. Shown here in the early 1950s, the train is carrying a steel General American-Pfaudler car leased by Whiting, followed by a pair of Rutland can cars and several B&M can cars. *S.K. Bolton, Jr.*

2-23

The modern Gehl Cheese plant in Germantown, Wis., receives sugar and soybean oil in tank cars. *Bob Gallegos*

2-24

Creameries with bottling plants require a fleet of delivery trucks. This is a Borden truck from the early 1960s. *Jeff Wilson collection*

at the various towns along your layout. You can also model a more limited operation, perhaps having the milk traffic handled by a local passenger or freight train.

You can also model the service found in many areas of the country in which milk was simply handled in baggage-express cars. As the local train stops at a town, the morning's milk cans would be on a baggage cart ready for loading. Or, perhaps the cans aren't yet ready and a train has to be held until they're loaded. If you model a terminal, you can have an REA or creamery truck waiting to pick up the cans collected along the train's run.

Condensaries, butter plants, and cheese plants are located throughout the country and can be logically modeled on most layouts. These can handle two or three cars, or several more if you have room. Butter will go out in refrigerator cars, and loads of condensed and evaporated milk can go out in refrigerator cars or insulated or standard boxcars. Dried products, packed in barrels or sacks, can go out in boxcars.

Don't forget the inbound boxcar loads of packaging materials and containers. Larger plants might also receive inbound fuel loads (usually coal, sometimes LPG or fuel oil) for in-plant power

and boiler houses. (Smaller creameries usually received fuel by truck.)

Walthers recently released a complex of creamery structures in HO scale (Sterling Consolidated Dairy, No. 933-3799) that could easily represent a creamery with a bottling plant, a condensary, a butter plant, or a cheese factory. Laser-Art Structures has a wood creamery in HO (no. 181-680), the N Scale Architect offers one in N (10502), and many companies offer wood and brick structures that could represent creameries and related structures. Milk platforms are available in HO from Creative Model Associates (no. 1007) and JL Innovative Design (461).

If you model a creamery with a bottling plant, be sure to have several delivery trucks ready to set out on their route, **2-24**. Walthers offers a Divco milk truck in HO and Wiking has one in N. Classic Metal Works has refrigerated box trucks and semitrailers in HO and N, and Sheepscot makes kits for bulk milk tank trucks in HO.

Milk cars are available from Athearn (a 40-foot, ready-to-run wood Pfaudler car in N), Walthers (53-foot, wood-sided Pfaudler car in HO), and InterMountain (40-foot Pfaudler steel car in HO). Several milk cars have been imported in brass, and Funaro & Camerlengo has made several cast-resin kits for can cars and tank cars (including the Borden butter dish car) in HO scale.

Milk cans are a necessity for creamery scenes, and they're available from Berkshire Valley (O scale, nos. 514, 524), Creative Model Associates (HO, no. 1006), Evergreen Hill (HO, 659; O, 8059), The N Scale Architect (HO, 20042; N, 20011), Neal's N-Gauging Trains (N, 44), Scale Structures Ltd. (HO, 2277), Selley Finishing Touches (HO, 154), Sequoia (HO, 1011), and Period Miniatures (N, 2012).

Breweries

3-1

The brewing industry has a long history with railroads. Hauling cases and kegs of beer is only part of the story, as railroads also carry incoming grain, malt, and other raw materials, as well as barrels, bottles, and cans, **3-1**.

Some beer still travels by rail, but since the 1960s, trucks have claimed most of the traffic.

A gas tractor moves a pair of Union Refrigerator Transit ice-bunker refrigerator cars into position at the Miller Brewing Co. in Milwaukee in 1934. *Courtesy Miller Brewing Co. Archives*

31

U.S. breweries in operation

Year	Number of Breweries	Barrels brewed (in millions)
1810	132	0.2
1850	431	0.8
1860	1,200	1.6
1880	2,200	13.3
1900	1,700	39.4
1916	1,300	58.6
1935	750	45.2
1940	611	54.9
1950	407	88.8
1960	229	94.5
1970	142	134.6
1980	101	188.4
1990	286	201.7
2004	380	204.3

3-2 Excludes microbreweries

3-3

Miller and other brewers began shipping beer by rail in the mid-1800s. This 15-ton wood car was built in 1889. *David P. Morgan Library collection*

3-4

The Crescent Brewing Co., in Nampa, Idaho, was typical of small breweries in the early 1900s, featuring a multistory brick structure, tall smokestack, and rail siding. This view is from 1915. *Jeff Wilson collection*

Brewing history

Beer has been around for thousands of years before railroads came to be. Beer crossed the Atlantic with the first settlers, and by the end of the 1600s, many commercial breweries were in operation in the colonies. When the Civil War ended, the U.S. had more than 1,200 breweries, and by 1880, that number had grown to more than 2,200, **3-2**.

The reason so many breweries had sprung up was that beer became very popular. It was also very perishable, making it difficult to transport the finished product. Most early breweries were small operations, making enough beer for local use within a town or city.

After the 1880s, the number of breweries began to shrink, while the amount of beer brewed grew tremendously into the 1900s. This was largely due to railroads and the coming of trucks and improved roads. As with other industries, breweries grew larger as processes became mechanized and automated, making it cost-effective to brew in volume.

Louis Pasteur also deserves a great deal of credit. His discovery that heat would kill microbes in canned and bottled food led to pasteurization. For breweries, this meant that by pasteurizing bottled beer, it would stay fresh for months instead of days.

This led to the growth of regional brewing companies. The Miller Brewing Co. began shipping beer by rail before the Civil War, **3-3**. Anheuser-Busch (the E. Anheuser Co. until 1879) was the first company to pasteurize bottled beer, and in the late 1870s, Busch began a national-scale advertising campaign for its new brand, Budweiser. Shipping the beer required a fleet of refrigerator cars, and in 1878, the St. Louis Refrigerator Car Co. became Anheuser-Busch's first subsidiary.

Other brewers that began pushing their product nationally were Milwaukee's Pabst and Schlitz. These larger companies became known as shipping breweries, or shippers.

More than a thousand local brewers still operated into the 1910s, **3-4**. By 1900, the average brewery in the U.S. turned out 23,000 barrels of beer a year, but by then, the larger shippers were making more than a million: Pabst was the first to top the million mark, in 1893, followed by Anheuser-Busch in 1901, then Schlitz.

A significant change was coming, in the form of the 18th Amendment: Prohibition. It prohibited the manufacture and consumption of beverages containing more than one-half of one percent alcohol. The amendment was ratified in 1919 and took effect January 16, 1920.

Many breweries – especially small ones – simply shut their doors. Others, including the large shippers, managed to stay in business selling products such as non-alcoholic beer, soft drinks, rubbing alcohol, candy, and dairy products. Some made malt syrup, which – although used in baking – was a key ingredient for home-brewed beer, **3-5**.

In April 1933, beer with a 3.2 percent alcohol content became legal again, and the 21st Amendment (repealing the 18th) took effect in December. About 300 breweries survived Prohibition, with about 700 operating by 1934.

3-5

During Prohibition, many brewing companies made soft drinks and malt syrup, products highlighted on this early 1930s Miller billboard reefer. *Courtesy Miller Brewing Co. Archives*

3-6

The Coors brewery in Golden, Colo., was served by multiple tracks from a curved spur off the main line at the top of the photo. In this early 1930s scene, several Coors billboard reefers are visible along with Pacific Fruit Express cars. *Hol Wagner collection*

The trend toward larger breweries continued, and the large shippers now had less competition from the hundreds of smaller breweries that had gone out of business. For example, the Coors brewery in Golden, Colo., became a large regional brewer in the 1930s, **3-6**. Improved highways and rail service also made it easier for large breweries to get their products into more local areas.

Top 10 brewing companies (millions of barrels)

Rank	1940		1960		1980		2004	
1	Anheuser-Busch	(2.5)	Anheuser-Busch	(8.4)	Anheuser-Busch	(50.2)	Anheuser-Busch	(103.0)
2	Pabst	(1.6)	Schlitz	(5.7)	Miller	(37.3)	Miller	(38.6)
3	Schlitz	(1.5)	Falstaff	(4.9)	Pabst	(15.1)	Coors	(22.4)
4	Schaefer	(1.4)	Carling	(4.8)	Schlitz	(14.9)	Pabst	(7.3)
5	Ballantine	(1.3)	Pabst	(4.7)	Coors	(13.8)	Yuengling	(1.4)
6	Ruppert	(1.2)	Ballantine	(4.4)	Heileman	(13.2)	City Brewery	(1.3)
7	Falstaff	(0.7)	Hamm's	(3.9)	Stroh	(6.2)	Boston Beer Co.	(1.3)
8	Liebmann	(0.7)	Schaefer	(3.2)	Olympia	(6.1)	Latrobe	(1.1)
9	Hamm's	(0.6)	Liebmann	(2.9)	Falstaff	(3.9)	High Falls	(0.7)
10	Blatz	(0.6)	Miller	(2.3)	Schmidt	(3.6)	Sierra Nevada	(0.6)

3-7

When Prohibition started, breweries had to decide whether Prohibition was likely to be temporary or permanent. Those breweries that thought it was temporary continued to upgrade their machinery, bottling equipment, and processes while making other products – giving themselves a competitive advantage when Prohibition ended.

Along with expanding existing facilities, major brewing companies began buying other breweries or building additional facilities. Among the earliest to do so were Schlitz, which bought a Brooklyn brewery in 1949, and Anheuser-Busch, which built a new plant in Newark, N. J., in 1951.

In 1950, the country's top 10 brewers represented 38 percent of total production. By 1970, that figure was up to 69 percent, and today, the top four brewers account for 93 percent, **3-7**.

The number of brewers continued to shrink through the 1980s. Anheuser-Busch topped 10 million barrels in 1966, and others, including Coors, also grew, **3-8**. But many local and regional breweries closed, with some selling their brands to larger breweries. The number of active breweries reached a low of 80 (operated by 51 companies) in 1983. Even large brewers had their problems, as Schlitz closed its doors (with its brand bought by Stroh) in 1981.

However, the 1980s marked a turnaround in brewing, as the decade saw the growth of microbreweries that turn out fewer than 100,000 barrels annually. With more than 500 microbreweries in operation, these now account for about 5 percent of the market.

Raw materials

While the process of making beer became largely mechanized during the 20th century, the basic processes and ingredients haven't changed for hundreds of years. Beer is a natural product that has

The Coors brewery at Golden, Colo., is the country's largest. Here a pair of Coors plant switchers move cars at the complex in 1997. The coal cars in the foreground are for the brewery's power plant. *Matt Van Hattem*

3-9

Malting companies resemble grain elevators. Multiple covered tracks handle unloading for incoming loads of barley as well as the loading of outbound cars with malt. This is an Anheuser-Busch plant in Manitowoc, Wis., in 1992. *Richard Cecil*

3-10

The rear of the Manitowoc malting plant shows the elevators (at right), with the buildings housing the steeping tanks and kiln at left. Note the ductwork on the roof and the elevators painted to resemble cans and a bottle of Budweiser. *Richard Cecil*

just four key ingredients: water, malted barley, hops, and yeast.

Malted barley is the prime ingredient. Special types of barley are grown for beer making. The majority of six-row barley (described by the number of rows of flowers on the plant) used in brewing comes from the Northwest, Minnesota, the Dakotas, and western Canada. Some two- and four-row barley is imported from France and Germany.

Barley is sometimes malted by the brewery, but most often by a separate company or at another plant, **3-9** and **3-10**. Malting breaks down barley's starch and enzymes, resulting in simple sugars that feed the yeast during the brewing process, which produces alcohol.

Malt is key to a beer's flavor. To make malt, the barley is soaked in water for several days in steeping tanks. It is then drained and kept at a constant temperature in ventilated germination compartments or malting boxes. After a few days, the husk opens and the barley starts to sprout. This is known as green malt. The green malt is then heated and dried in a kiln for two to three days to stop the germination process.

The resulting malt depends on the barley used, drying temperature, and time. The longer it is

3-11

Large breweries comprise a variety of buildings, with modern concrete and steel structures often added to older brick complexes. The Schlitz trailers haul finished products to trackside, where they are loaded in railcars. This is along the Milwaukee Road's Beer Line branch in Milwaukee in 1980. *Richard Cecil*

3-12

A Milwaukee Road switcher shoves a cut of grain boxcars toward the Schlitz elevators in Milwaukee in the 1950s. A Pabst elevator is at right. *Wallace W. Abbey*

3-13 A covered hopper is being unloaded at the Pabst elevator in Milwaukee (at right). Spent grain can be loaded out at the track at left. The open hopper car in the distance is being loaded with cullet (broken glass). *George Drury*

malted, the darker it becomes – which affects a beer's color and flavor.

Finished brewing malt can be stored for several months until needed. Brewing malt is then shipped to a brewery, which mills it for the brewing process.

Hops are another key ingredient, as they control a beer's bitterness and some of its flavor and aroma. Hops, which are the flowers of the hop vine, look somewhat like green/yellow pine cones with layers of green scales. Domestic hops for brewing come mainly from Washington state, while some breweries import hops from Germany, England, or other countries. Each type of hops has its own characteristics.

Fresh hops have a high moisture content, and once picked, they must be kept cool to prevent spoiling. Some brewers use hops that have been dried and pelletized, which don't require refrigeration.

Special yeast converts the grain's sugar into alcohol and carbon dioxide, and the type of yeast also influences flavor. To get consistent results for a beer formula, brewers grow yeast in-house from known starting cells. Yeasts either rise to the top of the beer during fermentation (top fermenting), which results in ale, or ferment at the bottom, resulting in lager.

Brewing process

Specific brewing processes vary by brewery, but follow the same basic steps. At the mash tun (tub), ground malt is added to hot water that has been sterilized and filtered. The malt proteins break down with heat as the mixture is stirred, and the malt starches transform into fermentable sugars. Other grain can be added in the form of cereal mash, typically using either corn grits or rice. Some brewers add corn syrup to increase the sugar content.

This mixture then goes to the lauter tun, where liquid is drained from the bottom, then brought back to the top to be filtered through the spent grain. Once this process is complete, the liquid (now called wort) is separated from the spent grain, which is saved and sold as livestock feed.

The wort is funneled into brew kettles (also called brewing tanks or, in older breweries, coppers), where hops are added and the wort is brought to a boil. The initial hops added are called starting hops, and they control the beer's bitterness. Hops added at the end of the boil (finishing hops) influence the flavor.

The finished wort is then drained and cooled to about 50 degrees for lager or 70 degrees for ale. It then goes to a starter tank, where yeast is added to begin the fermentation process.

Fermentation takes about a week or two for ale and up to five or six weeks for lager. Fermenting generates heat, so these tanks are cooled to keep them at the proper temperature.

The yeast eventually settles out as the sugars are spent. The liquid, now green beer, goes to pressurized aging or lagering tanks for secondary fermentation and maturing. The alcohol and carbon dioxide levels increase, and the beer's flavor improves.

Once the beer has been aged, which can take several weeks, it is filtered to remove any traces of yeast or hops. It is then ready for sale, and goes either to the racking room to be placed into kegs or to the bottling room where cans and bottles are filled.

Packaged beer is generally heat pasteurized, then labeled (bottles) and placed in cartons. Kegs aren't pasteurized, so they must be kept cool in storage and shipping.

The brewery

Large breweries tended to grow in areas with significant German populations. Milwaukee,

Grain boxcars were emptied by removing the grain doors and shoveling the grain into hoppers next to the tracks. From there the grain or malt was conveyed to the elevators. *Wallace W. Abbey*

This long, low building was the bottling house at the Neuweiler Brewery in Allentown, Pa. *Jet Lowe, Historic American Engineering Record*

Forklifts came into popular use for loading cars in the 1960s. This is an interior loading area at Miller's Milwaukee brewery in 1964. *Courtesy Miller Brewing Co. Archives*

Workers transfer cardboard cases of bottled Blatz beer from a truck to an insulated boxcar in the late 1950s. Several breweries on Milwaukee's beer line hauled beer to team tracks and loaded reefers there. *Wallace W. Abbey*

These tracks at the Schlitz Milwaukee brewery were used for incoming cars of corn syrup (a tank car is in the shadows at left) and for outgoing loads of spent grain at right. *George Drury*

Storage tanks at the G. Heileman brewery in La Crosse, Wis., were painted to represent a six-pack of the firm's Old Style beer cans. *Richard Cecil*

St. Louis, Boston, Brooklyn, and Philadelphia were all important brewing cities. Access to ample supplies of water and (before mechanical refrigeration) ice were also important.

A brewery's size is based on output and how its processes are divided. A small brewery might have two or three buildings, **3-4**, while a million-barrel-plus shipping brewery will take up several city blocks, **3-11**.

The brewery itself houses the brewing operations, from mixing mash through lagering, and is the largest building in the complex, often standing several stories tall. The brewing process begins on the top floor, with gravity aiding the process through the brewing

stages. Large breweries often have multiple brewing buildings, depending upon the brewery's size or the number of different formulas being brewed. Offices may be included in this structure or located in their own building.

Grain and malt are stored in elevators, which look and operate like standard grain elevators, **3-12**. An elevator's size is based on the size of the brewery, **3-13**. Into the 1970s, when boxcars were commonly used for carrying grain, they were unloaded in protected areas, **3-14**. The grain or malt was pushed out of the car into a trackside bin and augered or conveyed to the elevator.

Bottling usually has its own building, **3-15**. Into the early

1900s, it was common for breweries to contract with outside vendors for bottling. A strange twist in federal tax laws until that time required beer to be placed in kegs before bottling, meaning bottling was often done off-site.

Warehouses are usually located near the bottling and racking houses, although, since it's a perishable product, breweries try to get beer loaded onto a truck or railcar as soon as possible. Warehouse space is also needed for cans, bottles, and packaging materials. Cars and trucks are often loaded inside, **3-16**. In another arrangement, trucks hauled the beer to team tracks, so the beer could be loaded into railcars, **3-17**.

This Union Refrigerator Transit billboard reefer was painted with a large Miller logo and lettering at the end of Prohibition in 1933. *Courtesy Miller Brewing Co. Archives*

Schlitz leased insulated boxcars from Dairy Shippers Despatch. Most DSDX cars were plain; this was the only one with the Schlitz logo. Note the return stenciling at left. *Richard Cecil*

Other railcars can be loaded or unloaded at outdoor or indoor tracks, depending upon the brewery, **3-18**.

Tanks of various shapes and sizes are located around breweries. These are sometimes decorated, **3-19**.

Barrels and kegs

Draft (keg) beer was the only type available until the mid-1800s, when bottles began to become available in limited numbers. Pasteurization led to an increase in bottled beer in the late 1800s, but 90 percent of beer in 1900 and 85 percent in 1919 were still sold in kegs.

This changed after Prohibition, as the production of keg beer dropped to 70 percent in 1935, 48 percent in 1940, and 12 percent by 1980. As home refrigeration became common, people simply brought bottles home rather than drinking at a bar.

Until the 1950s, beer was stored and shipped in wood barrels. A full barrel (31 gallons) is the standard measurement for beer, but the weight (300-plus pounds) made a full keg difficult to handle. Because of this, the half-barrel (15.5 gallon) and quarter-barrel (7.75 gallon) became common. Since keg beer had a limited shelf

life, the smaller sizes made it easier to consume before it spoiled.

Wood kegs were generally made with white oak staves and steel bands and were coated with pitch inside so that the beer didn't contact the wood. Kegs were returnable and were then cleaned and periodically re-pitched before refilling.

Coopering companies supplied breweries with wood kegs, and the breweries themselves devoted considerable time to keeping these kegs in good repair.

Following Prohibition, metal kegs were tried, but it took some

A Coors switcher pulls a couple of the company's insulated beer tank cars at Golden, Colo., in 2005. The building at right houses the coal dumper for the brewery's power plant. *Mike Danneman*

time before they were widely accepted. By the 1950s, aluminum and stainless steel kegs became the standard, as these were easier to clean and required less maintenance than wood.

Bottles and cans

Bottled beer was available in small quantities in the mid-1800s, but until the coming of pasteurization in the late 1870s, such beer had to be consumed within a few days of bottling. Also, into the 1890s, bottles were handmade, and thus expensive. Many capping methods had been tried, but none with great success until 1892, when the now-common crown cap was developed. The result was a tight seal that was inexpensive to apply. Coupled with new mass-produced bottles, bottled beer became much less expensive to produce.

Small breweries were more likely to continue shipping beer in kegs to their local bars, whereas bottles were more commonly used by large shipping breweries.

Bottles were generally shipped in wood cases, which – like the bottles – were returnable. These cases generally had hinged wood tops into the 1920s; later cases were open on top.

Cardboard cases became common following Prohibition, with heavier reinforced, reusable flip-top cases used for returnables and simple disposable corrugated cardboard boxes for later nonreturnable bottles.

The first no-deposit/no-return bottles appeared in 1939. They didn't become popular until the 1960s, but by 1980, disposables accounted for about two-thirds of bottles sold, and today returnable bottles are quite rare.

Returning kegs, bottles, and cases was a major headache – and expense. Canned beer, which first appeared in 1935, was one solution. Cans consumed fewer materials, had a lighter shipping weight, and took up less space than bottles. By 1941, cans accounted for 14 percent of packaged beer produced.

During World War II, cans were used only for beer produced for the military. Soldiers who became accustomed to canned beer during the war came to prefer it upon returning to civilian life, leading to an increase in canned beer production.

Canned beer surpassed bottled beer in sales in 1969, and today, cans account for 48 percent of all beer sold, compared to 43 percent for bottles and 9 percent for draft.

3-23

A string of grain cars, including one boxcar, waits near the Schlitz and Pabst elevators in Milwaukee in 1979. *Richard Cecil*

Railcars

From the late 1800s, beer was typically shipped in ice-bunker refrigerator cars, **3-20**. Kegs required refrigeration to stay fresh, and although bottles (and later cans) didn't need refrigeration, the insulated cars kept the beer from excessive heat or cold.

Breweries usually leased cars from refrigerator car companies, such as the Union Refrigerator Transit cars at the Miller brewery, **3-1** and **3-20**. Schlitz used Dairy Shippers Despatch cars, Anheuser-Busch owned St. Louis Car Co., and many breweries used Pacific Fruit Express or Fruit Growers Express cars.

Until 1934, many of these beer reefers were decorated in colorful billboard schemes, **3-20**. A 1934 Interstate Commerce Commission ruling on billboard refrigerator cars meant these cars were short-lived after Prohibition.

Another reason for the elimination of logos was for protection

against theft. Advised not to declare their cargo contents directly to thieves, companies switched to plain refrigerator cars, which became the rule from the mid-1930s on.

In the late 1950s and into the '60s, 50-foot insulated boxcars (AAR designation RBL) became common for hauling beer, **3-21**. These cars had much-improved insulation compared to earlier ice-bunker cars and could maintain their temperature for several days in transit. They also had a significantly greater capacity than the older 40-foot cars.

Coors currently operates a fleet of tank cars for hauling beer from its Golden, Colo., plant to bottling plants in the East. These insulated cars carry the beer under pressure, **3-22**.

Boxcars were used for hauling malt, corn, and other grain in bulk through the 1970s, **3-12** and **3-23**. Grain doors, nailed inside the door opening, kept the grain in place.

Covered hoppers for malt and grain began appearing in the late 1950s. They became the dominant method of hauling grain in the 1970s and were used exclusively from about 1980 on, **3-23**. Cars of 70-ton capacity were common into the mid-1960s, with 100-ton jumbo hoppers appearing after that. Malting companies sometimes operated their own cars, either owned or leased, **3-24**.

Fresh hops must be refrigerated, so they ride in refrigerator cars (ice or mechanical, depending on the era), **3-25**. Pelletized hops can come packaged in reefers or insulated boxcars. Cars in dedicated beer or hops service will be stenciled for return, **3-21**.

Some brewers began using corn syrup in the 1960s, **3-26**. The product is shipped in insulated, non-pressurized tank cars. Corn syrup is heavy (about 11.5 pounds per gallon), so cars hauling this product have historically been smaller than average. Modern

The Froedtert Malt Corporation leased this 5,400-cubic-foot PS-2CD covered hopper, built in 1974, from Pullman Leasing Co. *Pullman-Standard*

Fresh hops are carried in mechanical refrigerators, such as this Milwaukee Road car. Part of Pabst's old Milwaukee warehouse can be seen in the background in this late 1970s view. *Richard Cecil*

corn syrup cars are distinctive, **3-27**.

Standard boxcars were commonly used for incoming loads of bottles, cans, and packaging materials, although these often arrived by truck, especially if the suppliers were nearby.

Rail and truck operations

Small breweries typically had a single track or two serving them, and many small operations went to truck-only operations after World War II, **3-4**.

Most medium-to-large breweries have several tracks, each serving specific areas. The grain elevator is often a separate complex, and the bottling/canning building is also often separate from the main brewing building.

Breweries generally have their own power plants. Into the early 1900s, these were often coal-fired, with oil-fired boilers becoming more common from World War II on. These would require carloads of coal or tank cars of oil, often a few per week depending on the size of the brewery, **3-8**.

Loading usually takes place in a protected area, **3-16**. Sometimes this means having an apron over the loading dock, and sometimes the tracks extend into the building, **3-28**.

A train switching a brewery would have to come in, pull empty cars from the appropriate tracks, then set out loads (and empties for loading) at the appropriate spots. At a large brewery in a metro area, this could mean multiple switching jobs, with 40 or 50 cars a day being switched. A plant switcher could also be used, **3-8** and **3-22**.

A local brewery in a small town, on the other hand, would be likely to receive much of its raw materials by truck and might ship its products both by rail and truck.

Large breweries also have substantial truck operations. This can include inbound loads of hops, packaging materials, and sometimes grain. Outbound loads of beer will go in smaller trucks, for deliveries to area taverns, as well as in semis heading to distributors. Other outbound truck loads include spent grain (sold as livestock feed).

Anheuser-Busch, Coors, and Miller all still ship some beer by rail. As of 2003, Coors shipped about 150 cars a week, compared to 1,800 truckloads.

Before Prohibition, the destination for most beer shipped by rail was usually a brewery-owned depot, **3-29**. Brewers had extensive systems of branches, agents, and depots.

After Prohibition, restrictions on brewery ownership of taverns changed distribution patterns, with most now going to distributors and wholesalers.

Modeling

Many options exist to put all this information to use in modeling. First, realize that although we can come up with averages for many things, there was probably no such thing as a typical brewery.

To model a large brewery to scale would require the better part of a decent-size layout. Many small local breweries, which could be modeled to scale, unfortunately had only limited rail service or (especially after World War II) no rail service, relying entirely on trucks for receiving materials and delivering beer.

Larger regional and national breweries – those that served markets stretching out several hundred miles – were much more likely to use rail service for both raw materials and finished products.

Brewery size is expressed by barrelage – the number of barrels of beer produced per year. The historic big three brewers (Anheuser-Busch, Schlitz, and Pabst) each produced more than a million barrels per year at the turn of the 20th century.

By 1940, several other brewers had hit the million mark, and many others were over a half million, **3-7**.

This Crystal Car Lines car is typical of corn syrup cars used through the 1980s. It's shown at the Pearl Brewing Co., in San Antonio, Tex., in 1984. *Andy Sperandeo*

Modern corn syrup cars, such as this 19,000-gallon Trinity-built car, have a squatty look. *Jeff Wilson*

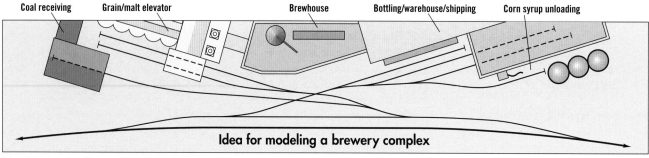

Coal receiving   Grain/malt elevator   Brewhouse   Bottling/warehouse/shipping   Corn syrup unloading

Idea for modeling a brewery complex

Model Railroader *illustration*

1 million barrel per year brewery in 1940

Incoming

Material	quantity	cars/year	cars/week
Barley malt	29,000,000 lbs	360 (40' box)	7
Corn	9,000,000 lbs	115 (40' box)	3
Corn syrup	2,500,000 lbs	36 (6,000 gal tank)	1
Hops	500,000 lbs	7 (40' box)	1
Bottles/cases	6,210,000 cases	4,700 (40' box or reefer)	90
Empty cans	966,000 cases	300	6
Empty kegs	960,000 kegs	3,000 (40' box or reefer)	58
Cardboard/packaging	840,000 lbs	12	1

Outgoing

Draft beer	960,000 1/2 bbl	3,200 (40' ice reefer)	62
Bottled beer	6,210,000 cases	4,500 (40' ice reefer*)	86
Canned beer	966,000 cases	400 (40' ice reefer*)	8

*Bottles and cans shipped in bunkerless or convertible-bunker cars if possible. Based on a split of 48 percent draft, 45 percent bottles, and 7 percent cans.

2 million barrel per year brewery in 1970

Incoming

Material	quantity	cars/year	cars/week
Barley malt	52,000,000 lbs	260 (100 ton covd hopper)	5
Corn	14,000,000 lbs	35 (100 ton covd hopper)	1
Corn syrup	5,600,000 lbs	48 (10,000 gal tank)	1
Hops	800,000 lbs	6 (70 ton reefer/covd hopper)	1
Bottles	9,384,000 cases	3,000 (50' box)	58
Cans	14,000,000 cases	2,500 (50' box)	48
Cardboard/packaging	13,600,000 lbs	97 (50' box)	2

Outgoing

Draft beer	600,000 1/2 bbls	1,000 (50' insul box)	19
Bottled beer	9,384,000 cases	3,475 (50' insul box)	67
Canned beer	14,076,000 cases	2,600 (50' insul box)	50

Weight of kegs, bottles, and cans

Kegs

Material	Size	Weight full (lbs)	Weight empty (lbs)
Wood	1/2 bbl	205	80
Wood	1/4 bbl	105	43
Steel (early)	1/2 bbl	190	65
Steel (early)	1/4 bbl	95	33
Stainless	1/2 bbl	150	30
Stainless	1/4 bbl	80	18

Package

Package	Weight full (lbs)
Wood case, 24 bottles	53
Cardboard case, 24 bottles	45
Cardboard case, 24 tin cans	28

Weights are approximate and will vary by brand and type of materials used.

Raw materials used to make 1 barrel of beer (lbs)

Ingredient	1949	1990
Malt	29.0	24.0
Corn	8.0	5.4
Rice	2.0	5.4
Hops	0.5	0.4
Corn syrup	2.5	2.9

Freight car capacity for barrels and cases

Car	Cases, bottles	Cases, cans	1/2 bbl kegs
40' reefer, bunkerless	1,400	2,400	—
40' reefer, bunkers in place	1,100	2,000	330
50' boxcar, insulated	2,700	5,400	590

3-28

The structures at the Coors Golden brewery form a tunnel for the company switcher as it heads to the indoor loading area to pick up loaded cars. *Matt Van Hattem*

3-29

A wagon and team pose in front of a Miller Brewing Co. depot in pre-Prohibition times. *Courtesy Miller Brewing Co. Archives*

How does that equate to railcars of beer and incoming raw materials? That depends upon the era you're modeling. The amount of malt and grain used per barrel of beer has dropped over the years, but the size of railcars has gone up.

For an estimate, determine the barrelage of your brewery, then use brewing statistics as a guide, **3-30**. The chart shows how much of each material is needed for each barrel of beer. Keep in mind that this is an average. For example, not all brewers added rice; those who did use rice, used more than noted in the average.

Scale these figures to match the proper percentage of draft, bottled, and canned beer. The listed totals show a summary of several sample breweries, **3-30**. Figure out how beer is shipped using the same charts – again, it will depend upon the cars used in your era.

Note that the numbers show all materials coming and going by rail, which would not have been the case. You'll have to determine for yourself how much your brewery ships and receives by truck.

Models of breweries have been made in HO by Design Preservation Models (no. 243-402), Heljan (322-807), and Vollmer (770-5609).

Many other structures are suitable, including the modular wall sections from DPM, Great West, Walthers, and others. Grain elevators from several manufacturers can be added to brewery complexes along with warehouse and bottling buildings.

Don't forget the details. Barrels are made by Finishing Touches, Grandt Line (HO wood), N Scale Architect (N wood), Preiser (HO wood, metal), and Scale Structures Ltd.

Modern beverage trucks are available from Boley in HO scale, and many other trucks are suitable for use as well.

Paper

4-1

Just think of all the paper products you touch in a day: books, magazines, paper plates, coffee filters, napkins, cardboard packaging, and masking tape. It's no surprise the worldwide paper market is huge – about 300 million tons per year. The U.S. supplies about 90 million of those tons, and Canada adds another 25 million tons.

The paper industry has long been, and continues to be, a vital customer for railroads, **4-1**. The raw materials, as well as the finished products, are well suited to rail transport. Along with outbound boxcars of paper products, mills receive inbound loads of wood chips, pulpwood, and various chemicals for production.

Regardless of the era your layout represents, you can model a portion of this industry or its traffic.

The Crown Vantage paper mill, near Berlin, N.H., looks like a very modelable mill. The boxcar in the foreground is carrying rolls of pulp. *Marty McGuirk*

FOUR

History

More than 5,000 years ago, people along the Nile River discovered that fiber from plants could be used to make a thin, flat material. Early papyrus (from which the word paper is derived) was more of a mat than today's paper.

The material we call paper today is a relatively new product, first made in China around the first century. It remained a closely guarded secret and didn't reach the rest of the world until several centuries later.

By the time Johannes Gutenberg used his new printing press to print his first Bible in 1456, paper was made primarily from cotton and linen rags. Rags were shredded and ground, mixed with water to form a pulp, and pressed onto a screen so the fibers from the fabric would bind together into a sheet. Papermaking was labor intensive: All the processes were completed by hand, with the paper made sheet by sheet.

In 1690, the first true paper mill in the colonies began production in Germantown, Pa. The mill made 100 pounds of paper a day, all by hand, using cotton and linen rags. Makers couldn't keep up with demand – there simply weren't enough rags available to make the amount of paper desired.

The first breakthrough, and the start of the modern paper industry, was the invention of the papermaking machine. In the late 1700s, Nicholas-Louis Robert developed a machine that made a seamless length of paper using a revolving wire mesh. The machine was perfected by brothers Henry and Sealy Fourdrinier, and the resulting machine that carried their name was patented in 1806. The machine revolutionized papermaking, as it eliminated much of the labor involved and allowed production on a much greater scale. By the 1840s, most paper mills had adopted machine technology, **4-2**.

During that period, enterprising individuals were developing ways of making paper using wood pulp instead of cloth. In the mid-

4-2

This 82-inch-wide Fourdrinier machine, built in 1887, was still in use in the 1970s at the Curtis Paper Mill in Newark, Del. *Historic American Engineering Record*

4-3

A Duluth & Northeastern locomotive switches two gondolas of pulpwood at the Northwest Paper Co. wood yard at Cloquet, Minn., in 1962. *John Gruber*

1860s, a mill in Quebec was the first to make wood pulp using a mechanical process, and within a few years, most mills were using wood pulp to make paper.

By the end of the 19th century, paper was widely used for newspapers and textbooks, and writing tablets were starting to displace personal slates in schools. By 1900, the U.S. produced 14,000 tons of paper per day.

Improved processes and larger equipment enabled greater production throughout the 20th century, and with the increase in paper products – such as packaging – U.S. mills now produce about 250,000 tons of paper products per day. Canada and the United States are both major paper producers, with large U.S. mills found in the Northeast, Southeast, and Upper Midwest. Smaller specialty mills are located across the country. There are currently more than 500 pulp and paper mills in the U.S. and about 160 in Canada.

Pulping

Papermaking comprises two distinct steps: making pulp and making paper. About 70 percent of mills are integrated and do both; however, many mills produce only pulp or paper. Let's start with a look at making pulp.

Almost all paper produced today is made from wood. (See the sidebar on cotton on page 53.) Raw materials include wood chips, sawdust, pulpwood, and recycled paper. Wood chips and sawdust are by-products from lumbering and lumber mills. Pulpwood encompasses trees and branches that are too small or unusable for lumber, **4-3**. Pulpwood generally makes up about half of incoming materials; the rest is wood chips, **4-4**, sawdust, and recycled paper.

Pulping is the process of separating the wood fibers, or cellulose, from the lignin that binds the fibers together. Several

Conveyors carry wood chips from the storage pile to the digester, the tall building at right. Pulp in liquid form is then piped from the tank's base (at right) to the papermaking building. The truck unloaders are at left. This is the Willamette Paper plant in Johnsonburg, Pa. *Jim Hediger*

Bales of kraft pulp, destined for use in making newsprint, are stockpiled at the Southland Paper Mill in Lufkin, Tex., in 1943. *Library of Congress (LC-USW361-836)*

processes are used, but all start by washing the chips (or grinding the pulpwood), screening, and sorting by type.

Different types of wood are used to make various kinds of paper. Softwood fibers are longer and stronger; hardwood fibers are shorter and weaker but result in a smoother paper finish. The two are often blended to get the desired type of pulp.

In mechanical or thermo-mechanical pulping, the chips go to a digester where they are impregnated with water and heated under high pressure until softened. They are then ground to release the fibers.

This process has a high fiber recovery rate (about 85–95 percent) but takes a great deal of energy, and the result is a low-brightness, low-strength pulp.

4-6

The long building at this Willamette plant holds the papermaking machine, with steam rising from the drying process. The track at right enters the building to the paper loading area. Chemicals and additives are unloaded and stored in the building and tanks next to the tracks at left. *Jim Hediger*

4-7

Multiple papermaking machines are in use at the Georgia-Pacific mill at Crossett, Ark. The dark-roofed building in the foreground is a tissue mill, with a paper mill in the long white building just above it. The pulp mills are in the center of the photo. Note the extensive trackage circling and entering the complex. *Georgia-Pacific*

4-8

The track next to the loading dock at the Gilbert Paper Co. in Menasha, Wis., is on a trestle. Paper mills require an abundant supply of water for operations. *Ted Schnepf collection*

This process is used most often to make newsprint.

In chemical and thermochemical pulping, fibers are separated by mixing the chips with various chemicals using pressure and/or high heat. The chemicals then dissolve the lignin and release the fibers.

There are two main chemical processes. In the sulfite process, introduced in the 1890s, chips are mixed with sulfuric acid and other chemicals under high pressure and temperature. The result is a light-colored, stable pulp that's easily bleached to get high-quality paper. Sulfite was the most common process into the 1930s.

The most commonly used chemical process today is known as the sulfate or kraft process (kraft means strong in German). In it, chips are mixed with caustic soda at high temperatures to separate the fiber. An advantage is that most chemicals in this process are recoverable. The resulting pulp is dark and strong but can be bleached to make lighter paper and paperboard. About 75 percent of U.S. pulp production uses the kraft process.

These processes recover only about 60 percent of the wood fiber, but the result is a brighter, stronger product than mechanical pulp. This is how most writing, magazine, and other high-quality paper is made.

Pulp for most printing paper must be brightened, or bleached, with a mix of chemicals. Chlorine was once the most common bleaching agent, but pollution problems (PCBs and other dioxins) led to the increased use of chlorine dioxide into the 1970s. Some plants have quit using chlorine in all forms; other chemicals used in bleaching include sodium hydroxide, hydrogen peroxide, and oxygen.

Pulp made at integrated mills is kept in liquid form, stored, and sent to the papermaking machine as needed via pipeline, **4-4**. Pulp is now made in a continuous process; early mills made pulp in batches.

Pulp made at one mill and sold for final manufacturing at another plant is known as market pulp. At pulp-only mills, the pulp is dried on screens in thin coats, cooled, and cut into sheets or rolled, **4-1**. Sheets are baled or palletized, **4-5**, and wrapped; the rolls and bales are then shipped to paper mills. Hundreds of grades of pulp are available, based on color, strength, and intended use.

Recycled (also called recovered) paper goes through a similar process. Incoming paper is shredded and mixed with water, then is de-inked. Chemicals were once used for de-inking, but modern processes use mainly water and detergents to remove ink, clay and other additives, and unusable (short) fibers. About 85 percent of fibers are recovered in the process.

Waste material from the pulping process (called black liquor) is removed, and much of it can be burned in various processes. Some of the residual material from de-inking can also be burned but must be placed in landfills or dealt with as toxic waste.

Modern mills have on-site wastewater treatment plants, and most water is recycled. Likewise, most chemicals are recovered and reused or treated. This wasn't always the case, and well into the 1900s, paper mills were a major source of air, solid, and liquid pollutants.

Papermaking

The paper machines used in modern paper mills are huge, often stretching 100 yards long and standing three or more stories tall. The basic papermaking process is similar to that used since the 1800s but with improved technology and on a much larger scale.

At an integrated mill, pulp is delivered directly from the pulper;

The warehouse at this mill on the Duluth & Northeastern features an indoor loading dock. *John Gruber*

Modern plants ship a lot of products on semitrailers, as shown by this early 1990s view of Woronoco (Mass.) Paper. *Jeff Wilson*

Paper products

Historically, most paper was used for printing, then writing. In the 1870s, cardboard, then corrugated cardboard, came into use as a shipping and packaging material. This segment grew substantially in the 1920s as cardboard boxes took over for wooden crates for many products.

Liquid packaging plants began appearing in the late 1940s as coated cardboard drink containers began taking the place of bottles for milk and other products. Other coated packaging included drinking cups and disposable paper products.

Tissue papers of various types came into wide use in the early 1900s, replacing the Montgomery Wards catalog as indoor plumbing became common even in rural areas. Facial tissue and paper towels also came into broad use, along with other hygiene products.

Today, paper products include newsprint and other printing and writing paper, cardboard for containers and packaging, food and product wrappers, and hundreds of other products.

4-11

Large wood chip gondolas have top-hinged end doors, allowing them to be dumped.
Burlington Northern

4-12

Tank car unloading areas at modern mills have concrete slabs and dikes to contain spills. Note the ground-level piping and hoses along the track for unloading. This car is carrying a sodium hydroxide solution. *Jim Hediger*

at a paper-only mill, pulp of various types is turned back into a liquid. The process starts on the wet end of the papermaking machine. At the blend chest, various types of pulp are combined, possibly with fillers such as clay or talc, to create the type of paper desired. The result of this blending is called furnish, which is diluted (about 99 percent water) and screened into the machine's headbox.

The furnish is then sprayed onto a large (up to 30 feet wide) continuous screen, called a wire, which can move more than 1,000 yards per minute. Dewatering begins immediately, using gravity, vacuum, and centrifugal force, and the fibers begin to bond together. The resulting fiber mat has a moisture content of about 75 percent. Now called a web, this fast-moving fiber mat is covered with felt and passes through rollers to squeeze more water from it.

Down to about 55 percent water, the web moves to the dry end of the machine. Here the web passes through a series of up to 100 steam-heated drying cylinders, each of which is four to five feet in diameter. This gets the moisture content under 5 percent. The paper then passes between large polished, cast-iron rollers that press the paper to the desired thickness.

Depending upon its final use, the paper can be treated (usually with starch) and coated. Materials used include mineral pigments – such as clay, chalk, or talc – and binders such as starch or protein. The paper is then dried again, using air or infrared afterdriers. High-gloss coated paper is then calendered, or ironed, to smooth it more.

Some paper undergoes further processing, such as embossing (often used with tissue and paper towel products).

The paper comes off the machine onto 30-foot-wide rolls,

with up to 40 miles of paper on a roll. The paper is then cut and re-rolled to the desired length and width.

Finishing mills – either at the same complex or at a different location – further prep the paper for use by cutting it to customer-specified dimensions, then encasing it in a vapor-barrier wrap. Other paper is cut into sheets, then palletized and stored or cut and packaged ready for shipment in boxes. Warehouses are climate-controlled to maintain proper moisture content.

Pulp and paper mills

Paper and pulp mills have grown significantly in size since the late 1800s. A newsprint papermaking machine in 1900 would produce a roll of paper, or web, about eight feet wide moving at 300 feet per minute; by the 1930s, machines turned out 24-foot-wide webs that moved more than 1,000 feet per minute. Buildings also grew to accommodate the equipment.

The tallest building at a mill is the digester, **4-4**. In modern plants, they can be up to 10 stories tall. Older mills had shorter buildings, but they often held multiple digesters and could sprawl out a bit.

The building holding the papermaking machine itself is usually long and comparatively low, **4-6**. Mills with multiple machines might have them in the same building or in different buildings, **4-7**.

A boiler power house, marked by a tall smokestack, **4-1**, holds the boilers and generating equipment to power the mill's machinery.

A warehouse – either a separate building or an extension of the papermaking building – is located at the dry end of the papermaking building, **4-8**. Tracks sometimes run alongside the warehouse, with a covered loading dock or doors, but tracks sometimes go into buildings, **4-6** and **4-9**. A truck

Tracks wind throughout the huge International Paper plant at Springfield, Fla. Here an Atlanta & St. Andrews Bay GP7 switches several chemical tank cars. *Scott Hartley*

Walkways, piping, and conveyors often cross over tracks at a mill, as at the Appleton (Wis.) Paper Co. One track goes into a building at right, while the paper loading dock at left is protected by an overhang. *Jim Hediger*

Paper from cotton

Some nonwood paper mills still exist, most of which use cotton plants for fiber. The raw material is generally linters, trimmings from the ginning process. Cotton mills produce high-quality specialty products such as fine writing and bond paper or decorative card and matboard.

4-15 Around 1960, Canadian National assigned several cars to newsprint service, marking them with yellow doors and stenciling. Note the newsprint note on the placard board. *Canadian National*

4-16 This 50-foot Railbox car is in paper service in this 1990s scene. *Jim Hediger*

4-17 Cars in dedicated service will have lading information and will often specify where the car is to be returned. *Donald Sims*

dock is usually located on the end or nonrail side, **4-10**.

Tanks are located near the pulp and papermaking buildings (often on roofs). A large concrete tank stands next to the digester to hold pulp. Other tanks vary in size and are generally steel, but many are lined to hold various chemicals. The tanks and buildings are connected by a maze of piping.

It takes about two tons of raw wood materials – plus chemicals and fillers – to make one ton of paper, so mills devote a lot of space to storing an ample supply of wood. Pulpwood, wood chips, and bales of pulp or recycled paper often come in by rail but also can arrive by truck.

The wood yard is a large area outside the plant where materials are stockpiled, **4-3**. Specialized cranes remove pulpwood from railcars and trucks. Pulpwood then goes into the barker, a large cylindrical enclosure that removes bark from pulpwood (the bark is burned in the plant). The pulpwood is chipped and stored in a large stockpile, **4-4**. Some plants don't chip pulpwood on-site, instead chipping off-site and transporting chips to the plant.

Modern wood chip cars are unloaded by end dumpers, **4-11**. Older wood chip cars could be unloaded by bottom dumping or rotary dumping.

Paper mills use a variety of chemicals and fillers in the pulping and papermaking processes, including chlorine, chlorine dioxide, sulfur and sulfuric acid, caustic soda, hydrochloric acid, and hydrogen peroxide. Fillers and coloring agents include starch, talc, kaolin clay, calcium carbonate, aluminum sulfate, rosin, and titanium dioxide.

Chemicals are often delivered in tank cars, **4-12**, but smaller plants receive many chemicals by tank truck or in steel drums. Powdered materials often arrive in covered hoppers or by truck.

These chemicals and fillers

4-18

Bales of pulp and recycled paper often arrive at mills in boxcars. *Jim Hediger*

4-19

Railroads often converted old hopper cars to wood chip service by adding side and end extensions, as on this Ashley, Drew & Northern car. *David P. Morgan Library collection*

were a major source of both water and air pollution around paper mills. Many rivers were contaminated with PCBs and other dioxins as a result of chlorine and other by-products that came under close scrutiny in the 1970s and after.

Because operations require millions of gallons of water per day, mills are almost always located next to lakes or rivers. Access to raw materials is also important, so mills tended to spring up where there was an ample supply of wood.

The size of plants is generally expressed in terms of the number of tons of paper produced per year or per day. Many early 20th century mills produced less than 100 tons per day; today's large mills with multiple papermaking machines turn out 2,500 tons per day. Many small mills having a single machine still exist and produce about 300 tons per day.

Because of the many types of pulp and the tremendous variety of paper products made, most mills specialize in a single type of paper. A newsprint mill, for example, only makes newsprint; its output is limited to large rolls destined for printing presses.

Mills require a tremendous amount of energy to operate, so they have their own power plants

4-20

Kansas City Southern bulkhead flats haul pulpwood to a mill near Shreveport, La., in 1977. *Jim Hediger*

4-21

Vertical logs placed at the ends of these gondolas expand their pulpwood capacity. *Jim Hediger*

Grades of paper

The five main types of paper products, from lowest to highest quality:

Paperboard (largest amount): cardboard boxes, gypsum paper

Packaging paper: grocery bags, craft paper, construction paper, packaging

Newsprint: newspaper printing

Tissue paper: paper towels, tissues, hygiene products

Printing and writing papers: office paper, writing paper, magazine and book paper

4-22

This ACF pressure tank car is typical of those used to haul liquid chlorine from the late 1930s through the 1970s. *Roy C. Meates*

4-23

This modern noninsulated tank car is in dedicated service carrying sulfuric acid. Chemical cars are stenciled with their lading and placarded. *Jeff Wilson*

to generate steam and electricity. These can use natural gas, coal (common through the steam era), fuel oil, or propane, which can be brought in by rail. Mills also burn by-products from the pulp and paper processes.

Some smaller nonintegrated mills might be served by just a couple of tracks, but larger operations – especially integrated mills – have tracks snaking throughout the plant, **4-13**, with specific duties for each track. Raw materials are delivered at one end of the plant, chemicals and fillers in the middle, and finished products at the other end, **4-14**.

Railcars

You'll find many types of railcars around paper mills. Boxcars are normally used for carrying finished paper in rolls, boxes, or pallets of sheets. The primary requirement of a paper-service car is that it be clean and dry. Forty-foot cars were typical into the 1960s, **4-15**, with 50-foot cars common after that, **4-16**. Plug-door cars are often used to provide a tight seal, although sliding-door cars are also used.

Paper cars are often in captive service and will often be stenciled with instructions to return them to a specific agent or city, **4-17**. Boxcars are also used for incoming loads of recycled paper or pulp, **4-18**. These are baled or, today, often palletized.

Modern wood chip cars are among the largest railcars in operation. These large gondolas, with a top-hinged door on one end, came into popular use in the 1960s, **4-11**. Prior to that, chips were often hauled in older gondola or hopper cars converted to wood-chip service by having taller sides and ends added to them, **4-19**. These came in many styles, depending upon the railroad that built them.

Pulpwood also arrives in a variety of car types. Purpose-designed bulkhead flatcars are often used, with logs cut into four-foot lengths and stacked in two rows angled slightly toward each other, **4-20**. Other cars have pulpwood cut to the full car width. Gondolas are often used in pulpwood service, with either temporary or permanent bulk-heads to increase capacity, **4-21**.

The many chemicals used by mills often arrive by rail, each in its own type of tank car. Chlorine was used extensively into the 1970s, carried by pressure tank cars from the late steam era through the 1970s, **4-22**. Non-insulated tankers are typical modern cars that carry sulfuric and other acids, **4-23**.

Tank cars also haul many powders in slurry form, especially since the 1980s. Examples are the kaolin clay car, **4-24**, and the titanium dioxide car, **4-25**.

Other powdered materials – such as starch, talc, and clays – are delivered to mills in covered hoppers. Before the wide use of covered hoppers in the 1950s, these materials were often shipped in sacks or in bulk in boxcars.

Fuel can be brought in by rail. Coal-fired plants typically received 50-ton coal hoppers through the steam era; many plants later switched to propane or fuel oil, which could arrive by tank cars.

Modeling

Paper mills, especially modern ones, are huge. It would be difficult to model one to scale in its entirety, so it's usually best to compress the most-interesting areas, model the loading and unloading areas, and represent the rest of the plant on the back-drop (or unmodeled off the front edge of the layout).

There are several ways to model a paper mill, capturing a plant's essence while providing ample rail operations in a reasonable area. This could be accomplished as a layout itself, **4-26**, or it could be a peninsula on an

Kaolin clay is often transported in slurry form in insulated tank cars like this one. *J. David Ingles*

Titanium dioxide slurry is delivered in distinctive funnel-flow style tank cars with an extreme wedge angle. *Jeff Wilson*

Paper transloading

Richard Cecil

Small docks like this one are often used when end customers, such as newspapers, aren't located on-line. Boxcars are spotted next to the raised platform, and then forklifts move rolls of paper from the cars to trucks for the trip to the customer. This center was on the Milwaukee Road in Milwaukee and served the *Milwaukee Journal*.

A small transloading center like this would be easy to add to a city scene on almost any layout.

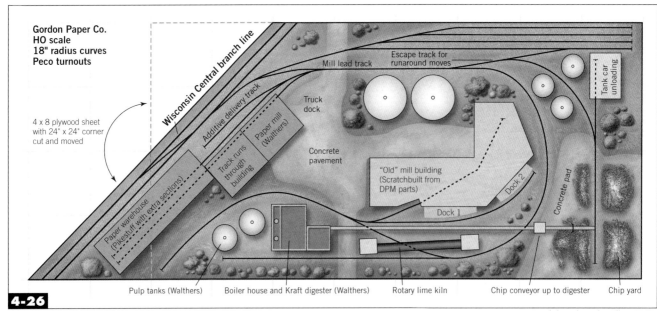

Model Railroader *illustration*

4-26

Gordon Paper Co.
HO scale
18" radius curves
Peco turnouts

Wisconsin Central branch line

Mill lead track

Escape track for runaround moves

Tank car unloading

Additive delivery track

Truck dock

Paper mill (Walthers)

Concrete pavement

"Old" mill building (Scratchbuilt from DPM parts)

Dock 2

Concrete pad

4 x 8 plywood sheet with 24" x 24" corner cut and moved

Track runs through building

Paper warehouse (Pikestuff with extra sections)

Dock 1

Pulp tanks (Walthers)   Boiler house and Kraft digester (Walthers)   Rotary lime kiln   Chip conveyor up to digester   Chip yard

around-the-walls layout. You could also rearrange it to fit along a backdrop by uncoiling the plant tracks and straightening them.

Mill buildings tend to be large, with brick construction used extensively into the mid-1900s and more modern structures of steel or concrete construction. Walthers has offered N and HO scale paper mill kits (Superior Paper Co.), including two buildings (digester and papermaking building) and several tanks, and the company also offered a pulpwood truck and other details.

Modular wall components are a good way to model the various structures. Brick versions are available from DPM (HO and N scales) and Walthers (HO), or concrete parts from Great West Models (HO). Pikestuff, Great West, Herpa, NuLine Structures, Walthers, and DPM offer HO and N scale industrial buildings that could easily represent various mill structures with little or no modification. Tanks, conveyors, piping, and other details are made by Plastruct, Pikestuff, Rix, Walthers, and others.

Appropriate boxcars are available from several manufacturers in HO and N scales. Almost any kind of 40- to 50-foot boxcar can be used depending upon era,

including HO scale 50-foot newsprint boxcars from Proto 1000. Atlas and Walthers have made pulpwood cars in HO (with loads from Walthers, Chooch, and Model Railstuff), and other bulkhead cars can also be used. Modern HO wood chip cars are made by LBF and N scale cars by DeLuxe, with loads from Model Railstuff.

Appropriate tank cars have been made by several manufacturers, including HO and N scale clay slurry tank cars from Atlas. Chlorine would be carried in pressure cars like the 1950s-era Atlas HO and N scale AC&F models and the Trix HO cars. Acid cars generally have no bottom outlets, and many have special linings. Chemical cars will carry placards and have stenciling for their commodity.

You can also model a plant on a smaller scale by modeling a pulp-only mill or a small specialty paper mill that buys market pulp to make its paper.

Rail operations
A small plant could be served by a local freight, with the train pulling loaded cars of paper, then spotting incoming loads and empties in their proper locations. Larger plants might have a dedicated

switch job, working the plant two or more times a day, moving cars to specific spots as needed.

Another option is to use a plant-owned switcher to do the work. A local freight will drop off cars, then the plant switcher moves cars as needed, spotting outgoing cars so the next local freight can pick them up.

Either way, potential switching moves are quite extensive. A day's work might involve pulling several loaded cars of paper products from one warehouse track and replacing them with a like number of empty cars, pulling empty pulp boxcars that have been unloaded from another track, and placing loaded cars of recycled paper or market pulp in their places.

It could also include pulling empty wood-chip cars from another track and replacing them with loads, pulling empty chemical tank cars and covered hoppers and spotting new loads, spotting fuel cars at the power house, and pulling a loaded tank car of waste chemical sludge.

Even if you don't model an on-line mill, you can simulate paper traffic, with individual cars or blocks of cars carrying paper products, pulpwood, wood chips, or chemicals.

Iron ore

5-1

Iron ore is the key raw material in making steel, and railroads have been involved in moving the material since the birth of the steel industry, **5-1**. The growth of iron ore mining and transportation was driven by the industrial revolution of the mid- to late 1800s, as demand for iron and then steel sharply increased. Iron ore was needed in large quantities to feed the blast furnaces of the many steel mills of the Ohio Valley.

Expansion of iron ore mining coincided with the growth of railroad lines across the upper Midwest. Discovery of vast ranges of ore, notably across Michigan's Upper Peninsula (U.P.) and northern Minnesota, drove the building of many railroads.

Although iron ore has been mined in other areas of the country, the vast majority (about 95 percent) comes from the Lake Superior area.

A pair of Lake Superior & Ishpeming GEs shove ore cars along the LS&I's concrete dock at Marquette, Mich., in October 2004. The ship is the *James R. Barker*, the third thousand-foot Great Lakes ship. Built in 1976, the *Barker* can carry 59,000 tons of iron ore. *John Leopard*

FIVE

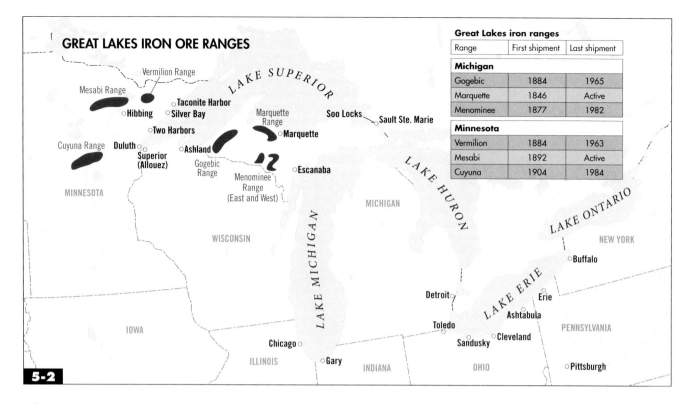

GREAT LAKES IRON ORE RANGES

Great Lakes iron ranges		
Range	First shipment	Last shipment
Michigan		
Gogebic	1884	1965
Marquette	1846	Active
Menominee	1877	1982
Minnesota		
Vermilion	1884	1963
Mesabi	1892	Active
Cuyuna	1904	1984

5-2

History

Most U.S. iron ore deposits are located in six iron ranges across northern Minnesota and Michigan's Upper Peninsula.

Ore discoveries generally followed an east-to-west pattern, first in Michigan with the first shipment of Marquette Range ore in 1846, then in Minnesota, **5-2**. In the early years, the ore was smelted near the mines. Charcoal that fired these small furnaces came from area hardwood forests, which were depleted quickly.

As the iron industry grew, coal became the primary fuel for furnaces, and it became more cost-effective to move the ore to the coal. As a result, steel-making centers developed on the southern Great Lakes as well as inland toward Appalachian coal fields.

A key development was the opening of the Soo Locks (at Sault Ste. Marie, Mich.) in 1855. This allowed ships to travel from Lake Superior to Lake Huron, from where they could proceed to Lake Michigan and south to Gary, Ind., or across Huron to the industrial centers along the south shores of Lake Erie.

The first Great Lakes ore dock was built at Marquette, Mich., in 1855. Wagons brought ore to the dock, where wheelbarrows transferred the ore to ships. At the time, the average ship carried about 250 tons of ore.

In 1857, the first railroad for moving ore was completed. The Iron Mountain Railroad (later part of the Duluth, South Shore & Atlantic) ran 16 miles from a mine to the harbor at Marquette. On the other side of the U.P., the Chicago & North Western opened a line from the Marquette Range to a dock at Escanaba in 1865. By the 1870s, the Marquette Range was producing more than a million tons of ore per year.

Large ore deposits were discovered in Minnesota beginning in the 1880s. Railroads carried this ore to docks in Duluth and Two Harbors, Minn., and Allouez (Superior), Wis. In 1880, Michigan produced 80 percent of the country's iron ore; Minnesota took the lead by 1900 thanks to the vast, easily accessible deposits in the Mesabi range, **5-3**.

Hundreds of mines produced ore in Minnesota and Michigan in the early 1900s, but the Depression slowed ore production to a trickle. It wasn't until the late 1930s and the approaching World War that the mines came back to life. By World War II, slightly more than a hundred mines were operating in Minnesota, with about 40 in Michigan.

By the late 1940s, the rich iron ore deposits were dwindling. In an effort to extend the lives of mines, pellet processing plants were built to process the raw ore and separate the iron. These pellets of taconite, hematite, and magnetite were soon being produced in large quantities and, by the 1970s, had supplanted raw ore. To be cost-effective, pellet plants need to operate on a huge scale – some producing more than 8 million tons of pellets per year – and need to run year-round.

Since the 1940s, mines have become larger, and there are far fewer of them. By 1981, 15 mines accounted for 90 percent of U.S. ore production. In 1991, 24 mines produced 56.8 million tons of usable ore (pellets and raw ore). By 1998, this had dropped to 12 mines, and in 2004, 10 mines turned out 54 million tons of ore, with eight mines accounting for 99 percent of production.

Today, the Lake Superior region still produces about 95 percent of the country's iron ore, with about 75 percent of that coming from Minnesota and 25 percent from Michigan, and railroads remain responsible for hauling the pellets to ore docks and steel mills.

Iron ore

Not all iron ore is alike. Ores from each range, and from different areas of each mine, have different chemical compositions. Ore has varying levels of iron and moisture content, as well as impurities such as phosphorous, silica, and manganese. Steel mills are particular about the specific makeup of each load of ore. To satisfy this need, suppliers customize the ore to each customer's specifications.

Through the first half of the 20th century, most ore mined was hematite. This ore has a high iron content (at least 50 percent), and could be used directly in steel mill blast furnaces. Called direct-shipping ore (or naturally existing ore), it was ready for shipment when dug from the ground – it only had to be crushed to size and loaded. Hematite is red, and its dust tends to coat everything.

Most high-grade hematite had been mined by the early 1950s. Concerns about running out of high-grade ore led to developments in refining that made it cost-effective to mine lower-grade ore and refine it to increase its iron content.

The process of refining ore to increase iron content is called beneficiation. Various processes have been used over the years, involving crushing, washing, and using magnets to separate the iron-bearing granules from waste.

Most ore mined today is taconite, an ore largely ignored until the 1950s. Taconite has a low iron content (15 to 35 percent), and is hard and abrasive. How-

5-3

A steam shovel loads iron ore directly into Duluth, Missabe & Northern railcars in Burt Mine on the Mesabi Range. Terracing makes it easier to remove materials. *Jeff Wilson collection*

ever, taconite is abundant, and experiments early in the 1900s showed that it was possible to draw usable ore from it. The first taconite plant in Minnesota opened in the 1920s, but it was a small operation, and a drop in the price of direct-shipping ore killed the operation after a few years.

Continued research and reduced availability of high-grade raw ore renewed interest in the process, and in 1948, Erie Mining opened a small taconite plant (200,000 tons per year) at Hoyt Lakes, Minn. The technology proved viable, and in 1957, the plant was expanded to produce 7.5 million tons per year. Other Minnesota taconite plants followed, such as the Ford-owned Fairlane plant of the Eveleth Taconite Co., **5-4**, with similar operations for refining hematite and magnetite in Michigan.

Making pellets

Pellet-making technologies have changed over the years, but the operation is basically the same.

Mined taconite is crushed, mixed with water, and finely ground. Magnetic separators remove iron-bearing particles from tailings. The concentrated iron mix (slurry) is mixed with bentonite clay and other materials and placed in large tumblers. The clay causes the material to bind together into small pellets.

The pellets are baked at 2,400 degrees to harden. The resulting pellets look like marbles (about ⅜" in diameter), have an iron content of about 65 percent, and are ready for the blast furnace. The pellets are purple-gray fresh from the kiln, but exposure to air and water turn the outer fine layer a reddish-rust. Pellets are then either stockpiled or loaded into ore cars.

Pellets from Michigan's U.P. are often referred to as taconite, but this is incorrect – it's either pelletized magnetite or hematite. Similar processes are used in beneficiating magnetite in Michigan; processing the nonmagnetic hematite requires a flotation separation process.

5-4

The Fairlane pellet plant, near Eveleth, Minn., was completed in 1965 and had an initial capacity of 1.6 million tons of pellets per year. Raw ore comes by rail from a mine 10 miles away, and pellets are shipped out. Loading and unloading take place on loop tracks (at left). *Eveleth Taconite Co.*

5-5

Steam rises from freshly loaded pellets in the ore cars of the train at left, while the train at right dumps its loads of raw ore from the Thunderbird Mine in this 1991 view. The Fairlane plant was expanded to a 2.3 million ton annual capacity in 1968. *John Leopard*

5-6

Tilden is one of two active Marquette Range mines. The massive Tilden processing plant ships about 7 million tons of pellets annually, hauled by Lake Superior & Ishpeming trains to the railroad's Marquette docks. Note the pellet stockpile at left. *John Leopard*

Regardless of the process, pellet plants are huge operations. They are usually located next to a mine, although some haul in raw ore by rail. The plants are marked by large buildings, stockpiled pellets, and loading conveyors, **5-4**, **5-5**, and **5-6**.

By 1965, more than half of Great Lakes ore was shipped in pellet form; by 1980, it was 80 percent, and today, it is virtually 100 percent. Pellets proved to be easier to ship, as they don't clump and stick in railcars and ship holds as does direct-shipping ore. They have less waste material, they melt well in blast furnaces, and their qualities can be controlled in the processing plants.

A variation on pellet operations developed in the 1980s. Flux pellets have limestone added to them, eliminating the need for steel mills to add limestone separately at the blast furnace. By 1990, flux pellets made up 39 percent of total pellet production. Some plants also ship out filter cake, concentrated ore that hasn't been processed into pellets.

Iron mines

Iron ore mines come in two types: Shaft (underground) and surface (open or open-pit). In an underground mine, a vertical shaft is dug into the ground, with horizontal tunnels (drifts) dug as the shaft reaches seams or pockets of ore.

Open-pit mining occurs where ore lies close to the surface. The top layer of earth (overburden) is scraped away to reveal the ore, which is then removed.

Early iron ore mines were basically open, shallow pits. However, by the 1870s, the easy-to-access, high-grade ore had been taken, and mines went underground. The technology is similar to coal mining, which is described in *The Model Railroader's Guide to Industries* (Kalmbach Books, 2004). Some shaft mines reached more than 3,000 feet deep.

5-7

The headframe of the abandoned Bruce Mine near Chisholm, Minn., is still standing. Ore was brought up on a skip from the shaft at right. The ore was dumped into the above-track loader at left, then into ore cars that traveled beneath the loader. *Jim Hediger*

5-8

Several cars have been loaded at the Armour Mine no. 2, near Crosby, Minn., in this 1912 view. The headframe is visible in the background, along with several buildings. Shaft mines have an ample stockpile of timber for bracing tunnels. *Minnesota Historical Society*

5-9

Abandoned mines tend to fill with water. You can clearly see the layers of red ore below the top layer of overburden. *Jim Hediger*

5-10

Large shovels and 240-ton trucks are now the standard method of removing ore from mines. *Jim Hediger*

Underground mining was dirty, dangerous work. Miners working in a drift would plant explosive charges to knock the ore loose. The ore was then loaded into small railcars pulled by electric engines. These were taken to an elevator (skip) in the main shaft, which raised them to the surface.

The top of the main shaft is marked by the headframe, which holds the pulleys and gearing for raising and lowering the skip, **5-7**. Adjoining buildings included a power or boiler house and a breaker and loader for crushing the ore and loading it into railroad cars, **5-8**. Other buildings included a dry house, where miners would change and shower, as well as office, storage, and maintenance buildings.

Below-ground mines had at least two main shafts: one with a cage for loading men and materials in and waste products out, and additional shafts with a skip for hauling ore from the mine. Cables from the power house operated the cages and skips.

Underground mining was expensive. The discovery of large surface ore deposits, especially in the Mesabi Range, led to the closing of many shaft mines in the early 1900s. The growing use of pellets spelled the end of the remaining shaft mines. The last below-ground mine in Michigan closed in 1979; the last in Minnesota had closed in 1967.

Surface mine operations are similar to coal mining. Large drills bore holes in the overburden, then explosive charges are planted below the surface. The explosions loosen the overburden so it can be stripped away by shovels or drag lines.

Open-pit mines are usually worked in terraces, **5-3**. This makes it easier for trucks or trains to gradually gain elevation as they climb into and out of the pit. By the 1930s, some of these mines were truly huge. The country's largest is Hull Rust Mine, near

5-11

5-12

Conveyors were often used to get ore from the pit to the surface. The long building is the beneficiation plant, with the spoil (waste) pile to the left. This is Hill Annex Mine, near Calumet, Minn., which closed in 1978. *Jim Hediger*

A Lake Superior & Ishpeming 2-8-0 eases a cut of ore cars back as a shovel loads cars from a trackside stockpile (background at left). The scene is at Cambria Mine near Negaunee, Mich., in August 1952. *Jim Scribbins*

Hibbing, Minn. Mined since 1895, Hull Rust now measures about three miles long, two miles wide, and more than 500 feet deep.

The deeper the mine, the more below-ground water seeps into it. Active mines have pumps working around the clock to keep water from collecting in the bottom. This is why abandoned surface mines quickly become lakes, **5-9**.

Once the overburden is removed, large shovels dig out the ore, loading it into trucks or railcars. Steam shovels, with some electric models, were used into the 1930s; diesel-electric shovels are also common.

Through the 1930s, it was common to use temporary railroads (steam or electric-powered) to haul ore out of the mine, **5-3**. Cars were loaded directly by shovels. The development of large dump trucks in the late 1930s made it easier to shift production to different areas of a mine, **5-10**, although some mines continued to use railroads into the 1990s. The size of both shovels and trucks grew steadily through the 1900s, and now 250-ton capacity trucks are common. Large conveyors are also used to get ore from the pit to surface level, **5-11**.

Once the ore is taken from the mine, it is stockpiled and loaded

5-13

Crude taconite is loaded into DM&IR ore cars at Thunderbird Mine in 1992. The train will soon make the 10-mile trip to the Fairlane pellet plant. *John Leopard*

5-14

Great Northern's no. 2 dock at Allouez, Wis., can fill two smaller ships at once. The gantry to the left of the diesel is a car shaker, which can travel along the dock. The ship, in this late 1950s view, is the *Cliffs Victory*, which could carry about 14,000 tons of ore. *Great Northern*

5-15

Below-track dock pockets are built on 12-foot centers. The long poles along the tracks are used by ore punchers to loosen any ore that sticks in the pocket. *David P. Morgan Library collection*

into railcars. This was once commonly done by shovels, **5-12**, but could also be done via overhead loaders, **5-13**.

Ore docks

Timber docks were the norm into the 1900s, and some remained in service into the 1960s. The first steel dock was built in 1909 by the Duluth & Iron Range at Two Harbors. It was 920 feet long, with 148 pockets, and could hold 43,000 tons of ore. Steel and concrete docks have been the standard since then.

Dock capacity continued to increase, as demand increased and ship size grew. The Duluth, Missabe & Northern built a 2,300-foot, 153,000-ton dock at Duluth in 1918, which remains in service.

Construction details vary by railroad and location, but docks share some basic features, **5-14**. Docks are tall, 70 to 80 feet above the waterline, to allow ships to come in under the loading pockets. To gain elevation, railroads often use tall trestle approaches to the docks themselves.

Each dock is divided into individual bins, or pockets, located in pairs (one for each side of the dock) below the tracks. Each dock has four tracks, so two tracks feed each pocket. Pockets are built on 12-foot centers, half an ore car's standard length of 24 feet, **5-15**.

Through the latter part of the 1900s, unloading trains was largely a manual operation. Engines would spot cars over the proper pockets, then workers would open the hopper gates to release the ore. A problem with natural ore was that it tended to stick inside the cars. Workers known as ore punchers would work atop or inside the cars, using poles to knock the ore loose, **5-16**. When working in the cars, they wore harnesses clipped to safety cables to keep from slipping.

The loading process was greatly simplified with the change to pellets, as they flow more readily out of the cars and dock pockets.

Large chutes stand upright at each pocket. To unload a pocket, the chute is lowered to the ship and the gate opened, **5-14**. Since the 1980s, some docks have been rebuilt to use conveyor belts to load ships, **5-17**. This lets the conveyor reach across to load both sides of the ship. Before that, with increasing ship sizes, ships sometimes had to back out and turn around to load the other side.

The advent of pellet operations, which generally run full-time 365 days a year, meant pellets must be stockpiled during the winter when the lake is closed to shipping. This can be done at the pellet plant, at the dock, or at both locations.

The modern Union Pacific dock at Escanaba, built in 1969, uses a rotary dumper. Cars are positioned by a mechanical arm and dumped three at a time. The dumper uses a system of conveyors to route the pellets directly to the loading pockets or to the stockpile next to the harbor.

The DM&IR added a similar system to its no. 6 dock at Duluth in 1965. Sixty-four pockets were modified, from which pellets can be conveyed to a stockpile with a 2.2 million ton capacity. The railroad also installed a similar system in 1977 at Two Harbors.

Ore train operations

There's more to iron ore operations than loading cars at a mine and hauling them to the docks. Natural ore varies a great deal in composition, and it was up to the railroads to blend it to the customers' specifications.

The process started at the mines, where solid trains of ore cars were assembled. Through the early 1900s, when mines were often smaller operations, trains could be made up of cars from several mines; larger mines would ship entire trains at once. Ore trains tended to be heavy and slow, from 50 to 125 loaded cars

5-16

Ore punchers worked on (and sometimes in) the cars to knock the ore loose so it would flow into the dock pocket. *John Vachon, Library of Congress (LC-USF34-063820-D)*

5-17

Part of DM&IR's Duluth dock no. 6 was rebuilt with conveyors for ship loading in 1983. Part of the facility's stockpile can be seen through the opening at left. The *Lemoyne* is a 730-foot ship with a capacity of 28,000 tons. *John Leopard*

5-18

Articulated steam locomotives were standard ore-hauling power into the 1950s. One of the DM&IR's class M-3 2-8-8-4s leads a loaded train through Alborn, Minn., in 1959. *Franklin A. King*

5-19

Cars of various grades of ore were sorted so they could be blended at the ore dock. This is the Great Northern's classification yard at Allouez, Wis., in the late 1950s. *Great Northern*

into the early 1900s to more than 200 cars in the mid- to late 1900s, depending upon the railroad and ruling grade, **5-18**.

Ore samples were taken as cars were loaded. These samples were analyzed and the results forwarded to classification yards, **5-19**, where cars were weighed, **5-20**, then sorted to match the specifications of each buyer. The cars were then reassembled into trains and forwarded to the docks. It was important that each car arrive at the proper pocket.

The advent of pellets greatly simplified train operations, eliminating the need for sampling and switching. Since the pellets were made to specification at the plant, solid trains of pellets from a single plant were moved directly to the docks when needed.

Weather could throw a wrench into dock operations, as natural ore tended to freeze in the cars in late fall or early spring. Railroads used several methods to heat cars. The first routed steam from stationary locomotives to pipes alongside several yard tracks. Steam lines could be connected to holes in the sides of cars or placed directly into the loads, **5-21**. The cars were then forwarded to the docks and the next cuts of cars brought in for heating.

Railroads also used infrared thawing sheds, **5-22**, where cuts of up to 36 cars could be moved for warming. Both methods were expensive and time-consuming, so only used when necessary.

Pellets eliminated the need for heating cars, as they are dry and flow easily. They're often shipped when still warm, making an interesting sight in the winter, with steam rising from the cars.

Empty cars are pulled from the docks once unloading is completed, and any cars needing repair are set out at the shops. Trains of empties go back to the mines or pellet factories, sometimes in trains longer than those with loads. Since pellets are

Ore cars roll across a weigh-in-motion scale on the DM&IR. Cars must be weighed before they head to the ore docks. The car is one of the Minnesota cars first built in 1936. *David P. Morgan Library collection*

Steam lines were sometimes placed directly into ore loads to thaw them, as here on the Great Northern. *Great Northern*

5-22

Thawing sheds used infrared heat to thaw ore loads. This is the interior of the Chicago & North Western shed at Escanaba, Mich. *Chicago & North Western*

usually hauled in modified cars with higher side extensions, empties must be sorted and directed as appropriate to a mine or pellet plant.

Train operations in natural-ore days were seasonal, following the Great Lakes shipping season (usually late March to mid-January). Rail operations shut down for the season, with steel mills having stockpiled enough ore to last until shipping resumed.

Railroads have also always shipped a portion of ore by all-rail routes, interchanged with connecting railroads. This could be done for various reasons, including shipments to non-Great Lakes or Ohio Valley steel mills or for winter moves when the lake is closed to shipping. These all-rail moves have increased since the 1980s, and today often feature run-through motive power.

Water-to-rail operations
In addition to rail movements from mine to ship, plenty of ore travels from ship to steel mill via rail. The Pennsylvania RR was the country's second-leading ore hauler into the 1960s, carrying about 33 percent of the ore that arrived at the Lake Erie ports of Cleveland, Ashtabula, and Erie. In addition, the Pennsy hauled a great deal of the ore imported to the United States via ports at Philadelphia and Baltimore. This ore was destined for the many steel mills along the railroad.

In ports such as Cleveland, ore was transferred from ship to rail using giant Hulett unloading shovels with 17-ton capacity buckets, **5-23**. The shovels dug ore from ships' holds and placed it into above-track loaders at the dock. Four tracks ran under the loader, which had a capacity of about 60 tons per minute, **5-24**.

Hulett unloaders first appeared in 1899. They greatly sped ship unloading and were in most lower Great Lakes ports by 1910. With the coming of self-unloading ships in the 1970s, Hulett unloaders were no longer needed, and the last one was retired in 1992.

The Pennsylvania, the New York Central, and other eastern roads used standard hopper cars for most ore traffic. (The Pennsy didn't acquire dedicated ore cars until buying 2,000 in 1960.) Hopper cars carried coal from Appalachian mines to Lake Erie ports. After unloading, the cars were filled with ore and headed back south and east to steel mills.

5-23

Giant Hulett unloaders empty the hold of an ore carrier at Cleveland in 1943. The hopper cars hauled coal to the port and carried the iron ore on their return trip east. *Jack Delano, Library of Congress (LC-USW361-666)*

5-24

Overhead bins at the Hulett unloaders drop the ore through chutes into waiting hopper cars. *Jack Delano, Library of Congress (LC-USW361-664)*

Because of the ore's density, hopper cars could only be partially loaded, **5-25**. From the ground, the cars looked empty; from above (a modeling view), they had a distinct appearance.

Additional related traffic

Coal, used for fuel at mines and processing plants, often comes by ship to docks next to the ore docks, and is hauled to the plants. Coal also sometimes arrives via all-rail routes, and is interchanged with connecting railroads, **5-26**.

Another back-haul is limestone, used to make flux pellets. Often brought in by self-unloading ships and hauled in ore cars, **5-27**, or carried in side-dump cars, it can also come in by all-rail routes.

The bentonite clay used in making pellets often arrives by rail, **5-28**. It is generally carried in covered hoppers. The quantity needed is not as significant as other raw materials; processing uses only about 16 pounds of clay per ton of pellets produced.

Ore cars

Ore cars, often called jennies, are specialized hopper cars. Iron ore is a heavy, dense material, weighing about 150 pounds per cubic foot (coal is about 80 pounds per cubic foot). Because of this, ore cars are much shorter than standard hopper cars.

Early ore cars were made of wood into the 1900s, **5-29**. Construction switched to steel around 1900, first with 50-ton cars, then with 70-ton cars by the 1920s. Ore cars look similar at a glance, but their designs differ by railroad and region.

Great Lakes ore cars have a coupled length of 24 feet to fit the spacing of pockets on ore docks. In 1936 and 1937, the first cars of a new style (Minnesota cars) were delivered, **5-20**, and became the car of choice among railroads with docks in Minnesota.

These cars were a foot wider (ranging from 10'-5" to 10'-9")

5-25 A pile of iron ore over each truck was enough to reach the weight limit of a standard hopper car. This is on the Pennsylvania RR at Erie, Pa. *Pennsylvania RR*

5-26 Two LS&I diesels shove a cut of coal hoppers into the coal dumper at Michigan's Empire Mine in 1987. The pellet loader is at left. *Eric Hirsimakei*

5-27 Several cars of crushed limestone, back-hauled from the ore docks, rest behind a cut of pellet loads at the Empire Mine. The Michigan ore cars have been given side extensions to haul pellets. *John Leopard*

71

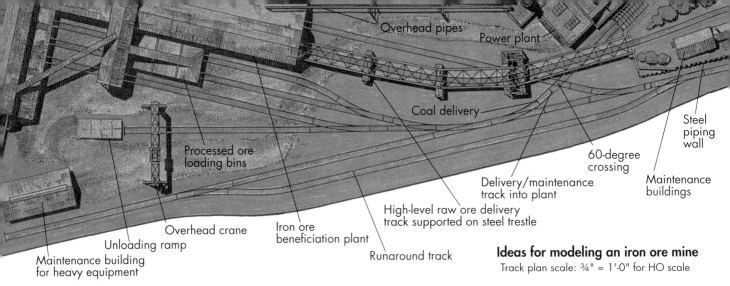

Overhead pipes

Power plant

Coal delivery

Processed ore
loading bins

60-degree
crossing

Steel
piping
wall

Delivery/maintenance
track into plant

Maintenance
buildings

High-level raw ore delivery
track supported on steel trestle

Overhead crane

Iron ore
beneficiation plant

Unloading ramp

Runaround track

Maintenance building
for heavy equipment

Ideas for modeling an iron ore mine
Track plan scale: ¾" = 1'-0" for HO scale

compared to the cars typically used in Michigan, **5-27**, and the Minnesota cars weren't as tall (10'-3" compared to 10'-9"). They took advantage of the wider track spacing on most Minnesota docks.

The DM&IR acquired almost 10,000 of these cars, and other railroads bought significant numbers. Many DM&IR cars were later sold to other railroads.

Ore-car spotting features include side shapes and patterns: some have rectangular sides, while others are angled. Most are riveted, but some are welded. Many were built with solid-

bearing trucks but received new roller-bearing trucks or had their old trucks rebuilt with roller bearings. Other differences include side sills, platforms and running boards, and brake hose locations.

In 1972, the DM&IR built its first miniquad car. These are four-car sets of older standard ore cars connected with drawbars and operated as a single car, **5-30**. They were designed especially for winter operation, cutting down on brake line air leakage by eliminating three air-hose couplings. The cars also limited slack action by

eliminating three coupler sets and were lighter than standard cars. They were marked by orange corner posts on the ends.

The miniquads also used a new brake system having two brake lines (hence, two sets of hoses on each end). This allowed engineers to make a brake-pipe reduction to apply brakes, while the other line could remain fully charged.

A common modification to ore cars was the 1960s addition of side extensions for hauling pellets. Since pellets aren't as dense as raw ore, cars can haul more of them before reaching their weight limit. Known as taconite cars (Minnesota) or pellet cars (Michigan), these cars have extensions ranging from 19 inches and 11.5 inches on early DM&IR cars to 9.75 inches on miniquad cars.

Modeling

Modeling an open-pit mine to scale would be difficult because of the sheer size involved. You could, however, model the buildings, loaders, and other equipment with the mine modeled out-of-sight or represented by a painting or photo on the backdrop.

Modeling a shaft mine would be easier in terms of space. You could model the headshaft, accompanying buildings, and the loader, along with loading and storage tracks.

5-28

An Erie Mining Baldwin switcher has spotted a covered hopper of bentonite clay at the Hoyt Lakes plant in June 1990. The car is a three-bay, 4,600-cubic-foot ACF Center Flow car.
John Leopard

Open-pit mine
painting on backdrop

No. 5 turnouts

Model Railroader illustration

It would be fascinating to model an ore dock. Walthers offered a steel ore dock model in HO (no. 933-3065), and AIM Products has conversion kits to turn that kit into a concrete dock. Resin Unlimited has announced HO kits for a Great Lakes ore carrier (no. 501) and the *Edmund Fitzgerald* (no. 503).

An ore unloading port would also be impressive. Walthers makes a large bridge crane (HO, 933-2906), designed for unloading coal from ships, that could be used for iron ore.

Ore cars have been offered by several manufacturers. Walthers has an HO model based on a DM&IR Minnesota car and a version with taconite extensions. Atlas has N and HO cars similar to a DM&IR car, and Model Die Casting has an HO model of a Michigan car following a Milwaukee Road prototype. Tichy has an HO turn-of-the-century wood car, and Bowser makes an HO model of a Pennsy G-39 ore car.

Chooch and Model Railstuff offer ore and taconite loads for most available models, and Woodland Scenics makes iron ore in several grades.

Even if you don't model an ore railroad, your layout can host part of an all-rail ore movement. Depending on the era, a train could feature motive power from the originating or destinating railroads.

5-29

The last Duluth, Missabe & Northern wood cars were 35-ton cars built by Pullman in 1900. *Franklin A. King collection*

5-30

The DM&IR's miniquads have four cars joined by drawbars. Advantages include lower weight and fewer air hose connections. *Harold Edmonson*

Historic iron ore railroads

The Erie Mining RR was known for its matched sets of EMD F9 diesels, shown here shoving a cut of loads at the dock in Taconite Harbor, Minn. *John Leopard*

Burlington Northern Santa Fe
BNSF (and **Burlington Northern**) predecessor **Great Northern** served several mines in the Mesabi and Vermilion ranges, hauling ore to a dock at Allouez, Wis. The **Northern Pacific** also served mines in the Cuyuna Range (under joint agreement with Soo Line), using the GN docks.

Chicago & North Western
A C&NW predecessor line was the first line from the Marquette Range to Escanaba in 1865, and the railroad also hauled ore from the Gogebic range to docks at Ashland and from the Menominee to Escanaba. Successor **Union Pacific** still hauls ore from Empire Mine to Escanaba.

Duluth, Missabe & Iron Range
The U.S. Steel-owned Missabe Road (spelled differently than the iron range) has historically hauled more ore than any other railroad. It was formed by the merger of the **Duluth, Missabe & Northern** (1892) and **Duluth & Iron Range** (1884) in 1938 and was purchased by **Canadian National** in 2004. It hauls ore from the Mesabi Range (and formerly the Vermilion Range) to docks at Two Harbors and Duluth.

Erie Mining RR
This railroad was built in 1957 to serve Erie Mining's new taconite plant at Hoyt Lakes, Minn., delivering pellets to Taconite Harbor. Mine trains delivered raw ore to the plant;

mainline trains hauled pellets to the dock. The line gained fame by operating matched sets of EMD F9s into the 1990s. The operation was bought by **LTV Steel Mining** in 1987 and was shut down in 2001.

Escanaba & Lake Superior
In 1901, the E&LS extended its trackage to Channing, Mich., to haul ore to the Milwaukee Road docks at Escanaba, an operation that lasted through 1936.

Lake Superior & Ishpeming
The Cleveland-Cliffs Iron Co. wanted its own railroad to haul ore from its mines near Ishpeming, so it built the LS&I in 1886. The LS&I still serves two mines in the Marquette Range, hauling ore to a dock at Escanaba.

Milwaukee Road
This road operated Menominee Range ore trains in a pool agreement with Chicago & North Western and also hauled Marquette ore to the dock at Escanaba.

Pennsylvania RR
The Pennsy hauled domestic ore from southern Great Lakes ports, as well as imported ore from Philadelphia or Baltimore, to inland steel mills at Pittsburgh, Johnstown, and other centers. Successors **Penn Central, Conrail,** and **CSX** continue to haul ore.

Reserve Mining RR
This railroad was built in 1955 to deliver ore from a mine near Babbitt, Minn., to a taconite pellet plant on Lake Superior at Silver Bay. The plant and railroad were shut down in 1986 but were reopened in 1990 by new owner **Cyprus Northshore**. In 1994, it was again sold and now operates as **Northshore Mining Co.**

Soo Line
Soo predecessor **Duluth, South Shore & Atlantic** was the first railroad to haul ore, to Marquette, and also hauled Gogebic Range ore to a dock at Ashland until 1965. The Soo also hauled ore in Minnesota in a joint operation with NP in hauling Cuyuna Range ore to the NP dock at Allouez.

Package and LCL traffic

6-1

In the days before UPS and FedEx became popular, it was the railroads' job to deliver packages, **6-1**. Whether it was a Christmas present from Aunt Edna or a lamp from Montgomery Ward, chances are it shipped via the Railway Express Agency (REA) or one of the railroads' own merchandise services.

Package and merchandise traffic, known as *LCL* (less-than-carload), was a major source of revenue for railroads through the mid-1900s. An extensive system of freight and transfer houses kept thousands of workers busy forwarding parcels.

LCL traffic is defined as a carload of freight from more than one shipper or destined for more than one receiver. This traffic fell into two basic categories. Standard LCL generally traveled in boxcars assigned to that service by the railroads. Express traffic was a premium service that warranted higher rates.

Workers carry a hand truck of less-than-carload items from a waycar boxcar to a depot on the Rutland at Florence, Vt., in 1958. *Jim Shaughnessy*

Milk cans and crates, just off-loaded from a train, share a cart at the Wells Fargo & Co./ Southern Express depot at Springfield, Mo., in August 1916. Government control of express services was less than two years away. *Lewis W. Hine, Library of Congress (C-DIG-nclc-05113)*

History

Getting into the business of delivering small packages has always been risky, from the days of stagecoaches to today. The problem is that if the volume of packages is too low, it's impossible to make money on them – especially if lots of pickups and deliveries must be made away from main routes. However, with a high volume of packages, the parcel business can be quite lucrative because the rate charge is much higher than for a bulk or carload shipment of the same weight.

The United States Post Office was founded in 1792, enabling people to send letters throughout the country. However, sending larger items, such as packages, required a private carrier or express service. Early pioneers in the express service included Wells Fargo, which provided New York-to-California package delivery in the early 1850s first by ship, then in the late 1850s by stagecoach.

Other express companies entered the business over the years, aided by the spread of U.S. railroads including the completion of the transcontinental railroad in 1869.

Railroads, as common carriers, were required to take packages and small shipments from businesses as well as individuals. Railroads consolidated packages going to common locations, transferred them at large freight houses, and from there redirected parcels to stations near the package destinations.

Parcel post service, begun in 1913, was an extension of the postal service that provided mailbox-to-mailbox delivery of small packages – although express companies saw it as unfair government-subsidized competition. Parcel post service was originally intended for shipping small packages (an 11-pound limit in 1913) but eventually accepted larger items (up to 70 pounds by 1931).

Express service commanded a premium rate and could cut days off long-distance shipments. By World War I, seven companies provided express services across the country, **6-2**. Private companies offering the service included Wells Fargo, Adams Express, American Express, and Southern Express; railroad-owned companies included Great Northern Express, Northern Express, and Western Express.

When the United States Railroad Administration (USRA) took control of the country's railroads during World War I, it also wanted control of the express companies that used railroads for shipping. The existing seven companies (with the exception of the portion of Southern Express that operated on the Southern Ry. and Gulf, Mobile & Ohio) were

Railway Express Agency opened a new terminal in Nashville, Tenn., in the early 1960s. The company's green delivery trucks were familiar sights in cities and towns throughout the country into the mid-1970s. *Railway Express Agency*

combined into a single operation, the American Railway Express Co., which coordinated all express shipments by rail. In the next few years, duplicate services were eliminated, many offices were closed, and routes and processes were consolidated and streamlined. This made it impossible for business to revert to individual companies after USRA control ended.

American Railway Express Co. continued operating until 1929, when all of its equipment and facilities were purchased by a group of 86 U.S. railroads. The new company, Railway Express Agency, became the sole provider of railroad express service in the country. (A lone holdout was the Southeastern Express Co., formed by the Southern Ry. in 1921, but it became part of REA in 1938.)

The agreement required each railroad to provide (and move) cars and provide terminal space.

Several cases of Kellogg's cereal are loaded onto a Cotton Belt truck for local delivery. Manufacturers often used railroad LCL service to deliver their products to businesses around the country. *David P. Morgan Library collection*

REA paid its own expenses, and any profit was divided among the railroads in proportion to total express traffic handled.

Green REA delivery trucks were familiar sights around the country until the company's end in the 1970s, **6-3**.

In the 1920s, many railroads began offering truck service to pick up and deliver packages for customers, **6-4**, but the improving national highway system and better and larger trucks meant that private carriers began eating into LCL business. The amount of LCL traffic carried by railroads peaked in 1923, and then slowly declined.

In an effort to keep this business, railroads in the 1930s and 1940s began running trains

6-5

The Southern Pacific used piggyback trailers on its *Overnight* LCL trains operating between Los Angeles and San Francisco in the 1950s. *Southern Pacific*

6-6

The Erie's 14th Street freight house in Chicago had seven tracks, each of which could hold about 15 40-foot boxcars. Trucks loaded and unloaded at the near platform. *Erie Railroad*

dedicated to LCL freight. Examples were Cotton Belt's *Blue Streak Merchandise*, Southern Pacific's *Overnight*, and New York Central's *Pacemaker*. Railroads also experimented with piggyback service and trailers for LCL traffic, **6-5**.

After a short boom in LCL following World War II, railroad business in package and express traffic rapidly declined. Freight forwarders and private carriers took a great deal of the business from the rails to the highways. Railroads hauled 24 million tons of LCL freight in 1946; this dropped to just 1.7 million tons by 1964, as railroads began to reduce – or completely eliminate – LCL service.

With the decline in passenger service into the 1960s, REA traffic also declined. The rail-based service could no longer match the efficiency of private package carriers such as United Parcel Service (UPS). UPS began as a local package delivery service in Seattle in the early 1900s and expanded service and increased business in the 1960s and 1970s.

A corporate reorganization wasn't enough, and REA (which had officially become REA Express in 1970) went out of business in 1975.

Railroads still haul a tremendous amount of LCL freight (now often termed LTL for less-than-load), but railroad employees no longer handle the packages. Instead, these packages ride in the trailers and containers of companies such as UPS.

Railroad LCL operations

The LCL (merchandise) business was lucrative for railroads, generating a lot of revenue compared to its weight. For example, in 1932, LCL traffic accounted for 10 percent of rail revenues but just 2.4 percent of the tonnage hauled. The service was, however, extremely labor-intensive. Tens of thousands of clerks, package handlers, and other workers

6-7

The Illinois Central's Dubuque, Iowa, freight house, shown here in 1916, is typical of those found in big towns and small cities. It has one covered track and two outdoor tracks. *Illinois Central*

6-8

An REA truck is backed up to the freight house at the New York, New Haven & Hartford combination station at Canaan, Conn. *Paul Larson*

6-9

Missouri Pacific's eye-catching blue and gray merchandise cars were rebuilt from older wood boxcars. The stencil at upper left reads, FOR MERCHANDISE LOADING ONLY BETWEEN M.P. LINES AND T&P FREIGHT HOUSES – DO NOT INTERCHANGE WITH OTHER LINES. *Missouri Pacific*

6-10

New York Central's Pacemaker boxcars were red and gray and were intended to stay on-line, running among system freight houses. *New York Central* sorted and processed the millions of packages handled by the railroads.

Each railroad had a network of freight houses and transfer stations that handled LCL freight. These ranged from huge buildings with multiple loading tracks in big cities, **6-6**, to small freight houses in smaller cities, **6-7**. Small towns were served by combination freight and passenger depots, which had attached storage rooms for LCL freight, **6-8**.

Packages collected from small-town stations were shipped to a larger freight house, where they were broken down, combined with other packages, and forwarded to another freight house closer to their destination. Packages proceeded this way until loaded on a car headed to their destination. Once packages arrived at their terminal, receivers either picked up their parcels or had them delivered by truck, **6-4**.

Large businesses, including mail-order companies such as Sears and Montgomery Ward, and manufacturers were big users of LCL services. A business loaded boxcars with boxes and packages heading to various destinations, then sent the cars to large freight or transfer houses. These trap or ferry cars were then unloaded and sorted into LCL cars based on their destinations (note the Kellogg's cereal boxes being loaded on a truck, **6-4**).

A variation of the ferry car was the line car. A company loaded a line car, then sent it along its route with multiple deliveries made from the car at various points along the route.

LCL cars weren't hauled on a random basis. Instead, cars operated on regular schedules between terminals, much like passenger trains. These schedules were published and distributed to shippers. Most of these routes operated every day (or daily except Sunday), although some operated less frequently (Monday-

6-11

Boxcars are parked in alignment with each other on parallel tracks, as here at Illinois Central's South Water Street freight house in Chicago. The blue flags on the car ends mean crews can't move the cars. Library of Congress *(LC-USW361-604)*

6-12

Steel bridge plates enable cars to be loaded on multiple parallel tracks. To work, all cars must be the same length. *Pacific Electric*

6-13

A powered cart hauls several carts of furniture to waiting boxcars at Seaboard Air Line's outdoor transfer facility in Hamlet, N.C., in the early 1950s. Several boxes of paint and varnish wait on another cart. *Seaboard Air Line*

6-14

The Atchison, Topeka & Santa Fe's Argentine, Kan., freight house featured automated cart tracks when it opened in 1961. *Santa Fe Ry.*

Wednesday-Friday and Tuesday-Thursday-Saturday were common variations).

Railroads routed cars between key freight houses on-line and also sent cars off-line to large freight houses or transfer stations of connecting railroads. Some routes rated a single car, while heavy-traffic routes required multiple cars.

Erie's Chicago 14th Street freight house was a typical large city freight house, **6-6**. As of 1930, this freight house loaded cars for 41 destinations (30 daily) and received cars from 32 stations (25 daily). Most points were on the Erie, but cars also came from, or were sent to, 14 other railroads in the Chicago area. The freight house received cars from Boston that originated on the Boston & Maine and the New Haven and also likely received ferry cars from local manufacturers.

Traffic and switching logistics

For the most part, LCL traffic traveled in standard 40-foot boxcars. Beginning in the 1930s, some railroads dedicated cars to LCL service. The Missouri Pacific painted cars in blue-and-gray *Eagle* colors, **6-9**, but probably the best-known was the New York Central's distinctive red-and-gray *Pacemaker* scheme of the 1940s, **6-10**. These specially marked cars were generally used only for on-line service. However, most railroads found it less restrictive to use cars that weren't in dedicated service.

At city freight houses, LCL cars were generally spotted in the early morning, having arrived during the night or early morning. Most large freight houses used multiple parallel tracks, **6-6** and **6-11**. Cars on the center tracks were reached by placing steel bridge plates between cars, **6-12**, making accurate spotting during switching critical. During unloading, cars were blue-flagged, **6-11**. Cars couldn't be moved while the flags were in place to protect the workers loading the cars.

Cars were loaded and unloaded using hand trucks or powered carts, **6-13**. As packages entered the freight house, they were marked for their destinations. Once inside, the packages were grouped for outbound cars. Some modern installations used automatic carts on belt lines to move parcels, **6-14**. Parcels for local delivery or truck transfer were routed to the appropriate truck dock, **6-15**.

Sometimes cars were completely filled, but more often than not, each car had a partial load. Packages were secured by boards, gates, or brackets to keep them from shifting in transit, **6-16**.

The cutoff time for loading was generally mid- to late afternoon, as most symbol freight trains left major terminals in the early evening. Cars carrying LCL freight were generally pulled by symbol freight trains, but some railroads operated LCL-only trains (see the sidebar on page 87).

At cutoff time, all the cars on a track were closed and each cut of cars was pulled out. At larger terminals, this string of cars was preblocked to eliminate excessive switching. The switcher took the cars to a yard, which was usually close by. Cars were then placed in their respective trains for departure. The loading and unloading schedule varied, depending upon the operations and train schedules of each railroad.

Although larger towns and cities warranted a dedicated boxcar, many small-town stations didn't generate enough LCL traffic to warrant an entire car. On these routes, LCL was often carried in a waycar, a boxcar loaded with the LCL freight destined for several stations along a line, **6-1**.

Carried in a local freight train (usually just ahead of the caboose), the waycar was spotted at the freight house or depot as the train arrived in town. If only a

A forklift loads a pallet of cigarettes onto a truck at the Kansas City Southern freight station in New Orleans. *Leon Trice*

This gate will keep the packages from shifting in transit. Train numbers and destinations are often chalked next to the door. *New York Central*

6-17

Railway Express Agency employees sort packages and place them on conveyors into baggage-express cars. *Railway Express Agency*

6-18

Bags of stored mail are loaded directly from a baggage car to a truck for a trip to the post office. *Ted Shrady*

6-19

REA acquired a fleet of 500 steel plug-door refrigerator cars, built by American Car & Foundry, in 1947 and 1948. *Railway Express Agency*

few packages were involved, the train could simply pause while it was unloaded.

Waycar service was common into the 1920s and found on some routes into the 1950s, but truck delivery began replacing waycars on many routes starting in the 1920s, **6-4**.

Express traffic

Unlike standard LCL traffic, which traveled in freight trains, express shipments rode in passenger trains. This meant increased rates (often double that of standard package rates), but shippers paid the price because express service could cut several days off transit times compared to standard LCL service.

Most REA freight was carried in railroad-owned baggage cars, more accurately called baggage-express cars. Although these cars indeed carried passengers' baggage, their main purpose was to handle REA's LCL traffic (and they carried RAILWAY EXPRESS AGENCY lettering), **6-17**, along with stored mail, **6-18**.

REA also operated an extensive fleet of its own, as well as leased, refrigerator cars, **6-19**, that were found in train consists throughout the country. These were also sometimes used in standard express service, **6-20**.

Items traveling as express included merchandise, individual parcels, time-sensitive goods, fragile merchandise, valuable items (such as securities, jewelry, gold, and silver), film and newsreels, news photos, machinery, and factory parts, **6-21**. Refrigerated express shipments included extremely perishable products (such as berries), flowers, highly desired products such as the first fruits of a season, and seafood.

REA handled mail between depots and post offices in many cities, and sometimes handled passenger baggage, typically between depots and hotels or between two depots. REA also had agreements at many locations (1,300 in 1940) for local pickup and delivery of railroads' own LCL traffic.

Operations followed a similar pattern to the railroads' own LCL services. REA trucks picked up packages from shippers and brought them to the local office, which was usually the railroad station. Passenger trains picked up packages, brought them to larger-city terminals, where they were consolidated, and forwarded them to a terminal near each package's destination, eventually loading them onto a delivery truck headed to the final destination.

Cars loaded and bound for a terminal were sealed, while cars used to deliver or pick up items at several stations along a route were manned by a messenger, who handled loading and unloading en route. Messenger cars were marked by a star on the side and equipped with a desk and toilet for use by the messenger.

Most depots sported signs indicating that they were REA offices, and REA's green trucks were a familiar sight around most depots, **6-8** and **6-22**. Incoming express packages were loaded onto baggage carts for transfer to the freight storage area or trucks, but packages and parcels were sometimes loaded directly from messenger cars into trucks, **6-23**.

As with railroad LCL cars, stations large enough to rate a full sealed car had that car spotted at the station or at a separate express building, **6-24**. But for many smaller offices, delivery was by messenger cars.

As with LCL waycars, many REA messenger car routes were eliminated and replaced by trucks starting in the 1940s. This was especially true after many passenger trains and routes were eliminated in the 1950s and 1960s.

Train operations varied by railroad. Express cars (known as head-end cars) were common sights on most passenger trains,

Boxes of Sears merchandise are being unloaded from this reefer, one of 300 converted Pullman troop sleepers operated by REA. *Paul Maximuke*

Insulated boxes of shrimp are loaded aboard a Louisville & Nashville baggage-express car at Biloxi, Miss. *Goodyear*

An REA truck, with driver at the rear door, stands ready as a Seaboard Air Line train pulls into Franklin, Va., on a foggy morning in 1962. *J. Parker Lamb Jr.*

6-23

Parcels were often transferred directly between express cars and trucks as in this 1947 scene on the Monon. Note the milk cans being loaded from the cart. *Linn H. Westcott*

6-24

Three baggage-express cars are spotted at REA's Nashville express station in 1962. *Railway Express Agency*

6-25

The Atlantic Coast Line's Waycross-Atlanta local carried a lot of head-end traffic, including four reefers and three express boxcars, in March 1954. *R.D. Sharpless*

but many railroads operated dedicated express/mail trains on heavy traffic routes, **6-25**. These trains, (with names such as *Fast Mail*) were often passenger trains in name only, with a string of baggage-express, express refrigerator, and mail cars trailed by perhaps a single coach or combine. Sealed express-baggage cars were often lettered for a neighboring railroad.

REA's decline

REA was a healthy organization through the 1940s. At its peak, it had 23,000 offices and 17,000 trucks. By 1939, REA handled air freight at 216 airports, contracting with several airlines. REA handled 231 million shipments in 1946, but by 1950, that number had dropped to 87 million. REA's ties to railroads put the company at a competitive disadvantage to trucking-based companies as the interstate highway system was built and larger trucks were introduced.

Starting in 1959, REA was allowed to route traffic via non-rail means with fewer restrictions, with more and more routes moving to trucks. The company began closing many of its small agencies at depots, opening what it called key-point terminals that used trucks to cover the territory of several closed offices.

However, service continued to suffer into the 1960s as railroad passenger service declined. Revenue was down, and railroads saw express traffic as more of a burden than it was worth. By 1969, only 10 percent of REA's revenue was from railroad operations, compared to 60 percent from trucks.

Although REA owned a large fleet of trucks, the company was unable to gain ICC approval to operate an extensive intercity trucking network through the 1960s, and REA's efforts to do so were fought by common-carrier truckers.

In 1969, REA was bought by five of its officers, and the company officially became REA Express in 1970 (it had used the new name since the early 1960s). However, increasing competition, a continued drop in traffic, and the elimination of most passenger trains doomed the company. REA hauled its last packages in 1975.

Modeling

Modeling LCL and express service could take many paths. If your layout includes a large metro area, you can include a large freight station or transfer house. Many of these lend themselves to typical model railroad design, with long, narrow buildings and multiple parallel tracks, **6-6** and **6-11**. These could be modeled against, or partially on, backdrops.

Switching these will provide operational challenges, with flurries of activity as inbound cars are placed in the morning, and another burst of activity in the late afternoon and evening as cuts of cars are pulled and quickly switched into dedicated trains or priority freights.

Operations in small towns and cities can be duplicated by switching at small to medium-sized freight houses or by placing LCL cars at the house tracks of combination depots.

You can also model waycar service on a branch line or secondary route by having a home-road boxcar placed near the caboose, pausing to simulate the time required to load and unload LCL parcels.

Express traffic can be modeled in much the same way, perhaps by having a passenger train set out a sealed car at a depot at a medium-sized town, or by having a train with a messenger car stop while delivering and picking up packages. Don't forget the REA trucks standing by.

You can also model an REA terminal, **6-3** and **6-24**, in a fairly

Merchandise name trains

New York Central

From the 1930s through the 1950s, several railroads operated trains dedicated to LCL traffic. Although most were not as famous as contemporary passenger trains, operating personnel knew all about them, and keeping them on schedules was a high priority of operation.

These merchandise trains included Cotton Belt's *Blue Streak Merchandise* (started in 1931) from St. Louis to points west, Southern Pacific's *Overnight* (1935) between Los Angeles and San Francisco, and New York Central's *Pacemaker* (1946), which originally ran between Buffalo and Boston and then system-wide. Other lesser-known trains include Lackawanna's *Pioneer* from Hoboken to East Buffalo, Chesapeake & Ohio's *Expediter* (1947) from Chicago to Newport News, and *Speedwest* (1951, reverse route). Some railroads ran unnamed merchandise-only trains, such as the Pennsylvania's, which carried the LCL symbol with the train number.

reasonable space. All you need is an arrival track or two, with a truck dock on the other side of the building.

You can model either LCL or express as through traffic, perhaps with a high-priority merchandise LCL train on a fast schedule, or with passenger trains carrying heavy loads of head-end traffic. Your passenger trains can include one or two express cars or refrigerator cars (or a complete express train) passing through, with sealed cars from foreign lines as part of the consist.

Walthers offers an HO kit for an REA transfer building (no.

933-3095), and many manufacturers offer models of depots and freight houses, several of which are based on specific prototypes.

Athearn and Walthers offer 50-foot wood express refrigerator cars in HO, and Walthers also has a converted troop sleeper in HO. Branchline and Walthers both offer REA 50-foot, steel plug-door express reefers, and Micro-Trains has one in N scale. Many companies offer baggage-express cars.

Use your imagination, and you'll be able to model what was once an important part of railroad traffic and operations.

Selected Bibliography

Coal customers

"Moving Coal Through Toledo," by Thomas N. Seay and Everett N. Young, *Chesapeake & Ohio Historical Magazine,* October 1998

When Coal Was King, by Louis Poliniak (Applied Arts Publishers, 1970)

Milk and dairy traffic

"Bennington County Co-operative Creamery," by Warren Dodgson. *Rutland Newsliner,* Summer 1992, p. 17

Land O'Lakes: Celebrating Tradition, Building the Future (Land O'Lakes, 1996)

"Milk Operations for Modelers, Part 1," by John Nehrich. *RailModel Journal,* November 1991, p. 44

"Milk Operations, Part 3: Private Owner Milk Cars in the Northeast," by John Nehrich. *RailModel Journal,* January 1993, p. 33

"Milk Trains, Milk Cars, and Creameries," by Chuck Yungkurth. *Railroad Model Craftsman,* June 1974, p. 26

"The Milk Trains: Transportation of Milk in New England," by Robert F. Cowan. *B&M Bulletin,* Winter 1977–78, p. 5; Spring 1978, p. 6

"Modeling the Milk Trains, Part 1: Midwest and East," by Robert Schleicher. *RailModel Journal,* August 2005, p. 10

"Modeling the Milk Trains, Part 2: 40-foot Pfaudler Milk Tank Cars," by Robert Schleicher. *RailModel Journal,* September 2005, p. 50

New York, Ontario & Western Railway: Milk Cans, Mixed Trains, and Motor Cars, by Robert E. Mohowski (Garrigues House Publishers, 1995)

"To Market by Rail: Milk Cars," by Chuck Yungkurth. *Railroad Model Craftsman,* February 1986, p. 89

"To Market by Rail: Privately Owned Milk Cars," by Chuck Yungkurth. *Railroad Model Craftsman,* March 1986, p. 85

Breweries

Beer and Brewing in America, by Morris Weeks Jr. (U.S. Brewers Foundation, 1949)

"The Beer Line," by Wallace W. Abbey. *Trains,* August 1952, p. 20

The Beer Book, by Will Anderson (The Pyne Press, 1973)

Brewed in America, by Stanley Baron (Little, Brown and Co., 1962)

Breweries of Wisconsin, by Jerry Apps (The University of Wisconsin Press, 1992)

Milwaukee Road's Beer Line, by Art Harnack (Milwaukee Road Historical Association Special Publication No. 5, 2003)

Paper

"A Mill You Can Model, Part 1: Action at a Paper Plant," by J. Emmons Lancaster. *RailModel Journal,* September 1997, p. 11

"Modeling a Modern Paper Mill, by Bernard Kempinski. *Model Railroader,* April 2002, p. 84

"Modeling the Paper Industry, Part 2: Railroad-Served Paper Sites," by J. Emmons Lancaster. *RailModel Journal,* September 1997, p. 17

"Newsprint Boxcars," by John Riddell. *Model Railroader,* June 2002, p. 70

"Papermaking and the Railroads," by Marty McGuirk. *Model Railroader,* October 1998, p. 100

"Papermaking and the Railroads Today," by Jim Hediger. *Model Railroader,* November 1998, p. 90

"A Rural Wood Chip Plant," by Kyle Lael. *Model Railroader,* October 2002, p. 96

Iron ore

Duluth, Missabe & Iron Range Railway, by John Leopard (MBI, 2005)

Great Northern Lines East, 2nd ed., by Patrick C. Dorin (Signature Press, 2001)

Iron Range Country (The Iron Range Resources and Rehabilitation Board and the State of Minnesota, 1979)

The Lake Superior Iron Ore Railroads, by Patrick C. Dorin (Superior, 1969)

The Missabe Road, by Frank A. King (Golden West Books, 1972)

"Burlington Northern: The Missabe Range," by John Leopard. *CTC Board,* January 1995, p. 28

"DM&IR's Miniquad Ore Cars," by Jim Hediger. *Model Railroader,* February 1976, p. 57

"Duluth, Missabe & Iron Range," by H.P. Scott. *Trains,* November 1942, p. 6

"Duluth, Missabe & Iron Range," Parts 1 and 2, by John Leopard. *CTC Board,* August 1992, p. 20; September 1992, p. 20

"Lake Superior & Ishpeming," by Mark Simonson. *CTC Board,* January 1990, p. 24

"LTV Steel," by John Leopard. *CTC Board,* December 1990, p. 26

"Minnesota Iron Mining," by Jim Hediger. *Model Railroader,* November 1992, p. 118

"The Minnesota Ore Car," by Pat Dorin and Jeff Koeller. *Mainline Modeler,* December 2003, p. 58

"Missabe Road Minnesota Ore Cars," by Patrick C. Dorin. *Model Railroader,* November 1992, p. 125

"Strong As Steel: Minnesota Iron Ore Keeps the Missabe Road Rolling," by Steve Glischinski. *Trains,* November 1992, p. 62

"Union Pacific: The Marquette Range," by John Leopard. *CTC Board,* May 1996, p. 49

"Who's the No. 2 Iron Ore Hauler? Pennsy!" by Bert Pennypacker. *Trains,* September 1962, p. 34

Package and LCL traffic

"Decline and Decay of REA," by Robert B. Shaw. *Trains,* July 1979, p. 22

"Erie's Western Terminus is Birthplace of Train 100," *Erie Railroad Magazine,* January 1960, p. 12

"A Modeler's History of Railway Express," by John J. Ryczkowski. *Model Railroading,* March 1987, p. 38

Model Railroading's Guide to Railway Express, by V.S. Roseman (Rocky Mountain Publishing, 1992)

"Moving the Packages," by Charles W. Bohi. *Chesapeake & Ohio Historical Magazine,* July/August 2005

Presenting the New REA Express Key-Point Terminal, Nashville (Railway Express Agency brochure)

"Profit Motive to the Rescue: REA," *Business Week,* December 8, 1962

"The Railway Express Agency," *Proceedings,* the Journal of the Pacific Railway Club, November 1940

"Railway Express Cars: An Update," by V. S. Roseman. *Model Railroading,* April 1991, p. 30

"When LCL Meant Fast Freight," by Charles W. Bohi. *Trains,* August 1993, p. 58

Additional resources

Car Builder's Cyclopedia (Simmons-Boardman), various issues

The Model Railroader's Guide to Freight Cars, by Jeff Wilson (Kalmbach Books, 2005)

The Official Railway Equipment Register, various issues